THE CITY

THEATRES OF ACCUMULATION

CITIES IN THE DEVELOPING WORLD

THEATRES OF ACCUMULATION

Studies in Asian and Latin American Urbanisation

WARWICK ARMSTRONG & T.G. MCGEE

Routledge
Taylor & Francis Group
LONDON AND NEW YORK

First published in 1985

This edition published in 2007 by
Routledge
2 Park Square, Milton Park, Abingdon, Oxon OX14 4RN
711 Third Avenue, New York, NY 10017

Transferred to Digital Printing 2007

Routledge is an imprint of the Taylor & Francis Group, an informa business

First issued in paperback 2013

British Library Cataloguing in Publication Data
A CIP catalogue record for this book
is available from the British Library

Theatres of Accumulation

ISBN13: 978-0-415-41739-6 (hardback)
ISBN13: 978-0-415-86032-1 (paperback)
ISBN13: 978-0-415-41318-3 (set)

Routledge Library Editions: The City

Theatres of Accumulation

Studies in Asian and Latin American urbanization

Warwick Armstrong
and T. G. McGee

Methuen
London and New York

First published in 1985 by
Methuen & Co. Ltd
11 New Fetter Lane, London EC4P 4EE

Published in the USA by
Methuen & Co.
in association with Methuen, Inc.
29 West 35th Street, New York NY 10001

Typeset by Scarborough Typesetting Services
and printed in Great Britain
at the University Press, Cambridge

British Library Cataloguing in Publication Data

Armstrong, Warwick
Theatres of accumulation: studies in Asian
and Latin American urbanization
1. Urbanization – Developing countries
I. Title II. McGee, T. G.
307.7′6′091724 HT153

ISBN 0–416–78570–0
ISBN 0–416–39800–6

Library of Congress Cataloging in Publication Data

Armstrong, Warwick.
Theatres of accumulation.
Bibliography: p.
Includes index.
1. Cities and towns – Developing countries – Growth.
2. Urbanization – Asia, Southeastern – Case Studies.
3. Urbanization – Latin America – Case studies.
4. Urban economics.
I. McGee, T. G. II. Title.
HT149.5.A76 1985 307.7′6′0959 85–11466

ISBN 0–416–78570–0
ISBN 0–416–39800–6 (pbk.)

Contents

List of figures

List of tables

Preface

Collaboration is an involved process. This book has grown out of twenty years of experience in the Third World. The authors first shared their experiences of the Third World while colleagues in the Department of Geography, University of Wellington, New Zealand, between 1965 and 1968. Our awareness was greatly heightened by the atmosphere created by Keith Buchanan, who was then one of the few English-speaking geographers to have opened new perspectives and sensitivities to the problems of the Third World. Since then our professional interests have led us separately into periods of residence and research in Asia and Latin America. The experience in two different continents has not, however, led to a divergence in our intellectual appreciation and awareness of the process of development.

On the contrary, it has resulted in a growing consciousness of the broad similarities of the process of capitalist expansion and accumulation at a global scale. The understanding of the specific historical manner in which these processes have affected societies or, indeed, regions and individual cities has provided us with a necessary insight into the interaction between the external and internal processes as they affect development.

This viewpoint is central to the structure of this book. In the first three chapters we advance a wide-ranging discussion of existing theoretical explanations of these processes in order to arrive at a synthesis of these approaches; this in our view is more valuable than favouring a particular ideological or theoretical stance. Our objective has been to avoid fashionable polemic – the raising of one theoretical position to pre-eminence, a practice, which, in our judgement, bedevils much of the development literature on the Third World. The process of research is not discontinuous but cumulative, and builds on the ideas of those who have previously contributed. Einstein's metaphor is clear:

> Creating a new theory is not like destroying an old barn and erecting a new skyscraper in its place. It is rather like climbing a mountain, gaining new and wider views, discovering unexpected connections between our starting point and its rich environment. But the point from which we started out still exists and can be seen, although it appears smaller and forms a tiny part of our broad view gained by the mastery of the way up. (cited in Zukav 1979, 19)

Taking this sense of interaction further, we find it necessary to state that our methodology is not based on a choice between either theory or practice; rather the relationship is a mutually reinforcing one in which both contribute to deepening our awareness of the nature of development, underdevelopment and social transformation. The general structures of grand theory are necessary to allow us to perceive the overarching trends of social evolution and change; the empirical case studies from Chapters 6 to 9 allow us to amend those perceptions in the light of smaller-scale and human reality.

Some readers may feel that this approach is unbalanced and that case studies do not help in the explanation of general theory. This is an old debate. In our judgement case studies illustrate the action of processes as they occur in specific and differentiated geographical locales in the global system. Case studies are snapshots which enable the 'freezing' of process in time and space to allow a deeper analysis of process. To this extent no case study can ever be typical; it can only illustrate certain aspects of the whole. What the Hong Kong hawker and Ecuadorian peasant have in common, despite their physical separation, is their relationship to capitalist expansion and accumulation.

Our title may mislead the reader into thinking that the book is concerned with the traditional subjects of urban structure and morphology. In fact we wish to avoid any reification of the city as the subject of independent study. Our objective has been rather to emphasize the role of the Third World city, first as the principal theatre of action for those decision makers concerned with the operation of capital, corporate business and the state. Then, secondly, the cities of Asia and Latin America, and especially the large metropolitan centres, act as the focal points of the capital accumulation process which stretches to the most remote rural parts of Third World societies. And it is in these cities that the major realignments in class relationships are emerging.

This process is often conventionally discussed in terms of so-called 'modernization' which in most cases is equated with conformity with western models. We have sought to break down this homogeneous concept by emphasizing the significant differences that occur within the spheres of production and consumption. It is in the sphere of *consumption*, most evident in the lifestyles of city populations, that a clear *convergence* is occurring throughout much of the Third World. There is a class character to this process, for it is among the upper and middle classes of the cities that convergence in lifestyles and possession of consumer amenities is most obvious. However, these patterns are now spreading to low income and rural groups (Wells 1977). In the sphere of production, in spite of considerable expansion of *capital* there is much more *divergence*, reflecting the increasing international division of labour. This divergence varies considerably according to the type of productive activity. In export manufactures the process of divergence is extreme, with only a few Third World countries participating. On the other hand, in import-substitution production (especially consumer goods) the process is less clear-cut, showing elements of convergence in management and technology.

While we have attempted to define our conceptual categories throughout the text, four terms are used so often it is useful to define them here.

Third World The meaning of this term is becoming more diffuse and some writers would prefer to discard it altogether. In this book we use the term to refer to those countries which, however differentiated from each other in their levels of pattern of development, have been subjects of penetration by the capitalist societies of Anglo-America, Western Europe and Japan, and their outliers in Australia and New Zealand.

Urban This is an almost undefinable term which is a required part of the language of contemporary social sciences. We have found it necessary to use a concept of 'urban' that is not linked to narrow settlement or administrative definitions, although clearly any statistical evaluation of urban growth relies in part upon such definitions. Rather, we equate the concept 'urban' with the idea of a facilitative framework of institutions occurring in a linked hierarchy of settlement, in which dominant classes, in particular, make decisions relating to the accumulation of capital. We have employed the term 'theatres of accumulation' to emphasize the class nature of the city's function. We realize that such a definition is broad in the extreme, but it at least enables us to escape from the crude kinds of 'urban' determinism which suggest that cities of themselves produce certain behavioural patterns and economic processes.

Capital accumulation This term, too, could be subject to volumes of analysis. We simply adopt the following working definition: at its crudest level capital can be regarded as comprising the assets arising from the production of goods and services. We see it in a more complex sense as representing the economic, social and political power of a class (capitalist). Capitalists derive their power (in the physical shape of money, plant and technology) through their capacity to extract surplus value from production through direct appropriation from wage labour, or other forms of surplus from trading and pricing activities, rent or taxation. These relationships are expressed in the brief statement that 'Capital is . . . by definition, value looking for accretion, for surplus-value' (Mandel 1976, 60). Accumulation is thus used in the sense of building up capital. The mechanisms of accumulation are more complex and are discussed in the main text, but, again, Mandel's brief summary is apposite: 'Under capitalism economic growth . . . appears in the form of accumulation of capital. The basic drive of the capitalist mode of production is the drive to accumulate capital. . . . It is essentially explained by competition, that is by the phenomenon of "various capitals"' (1976, 60).

Class We have chosen to use the term to describe the major socio-economic groupings of Third World societies, principally defined in terms of their relationship with capital and the process of accumulation. We do not think it is fruitful to proceed here with inferences on the likely development of class consciousness in these groups, given the wide range of possibilities for political action that exist.

Any book which involves the level of synthesis of this study clearly owes an intellectual debt to many; we are particularly grateful to our friends in Latin America and Asia who have tolerated our presence and encouraged our research. Grateful acknowledgement is made to Paul Jance for drafting figures 1.1, 1.2, 3.1, 3.2, 3.3, 3.4 and 9.1; to Sandy Pan and Sabrina Yan for typing much of the manuscript. We also wish to thank the World Bank for permission to reproduce Figures 1.1 and 1.2.

This book is dedicated by Terry McGee to Kamal, Mei Ling and Lean Heng, partners in 'the international division of labour'.

Warwick Armstrong wishes to thank, in particular, the many colleagues and friends who have contributed to the formation of his ideas over the past decade or so. Most would not recognize their contributions in the chapters which follow, but it is through the complex processes of discussion, testing ideas and working together in offices, classrooms, on farms, in factories and on the street that the ideas set out have gradually evolved.

It would be invidious to attempt to name particular individuals; instead by referring to the institutions and groups, they may recognize the role they have played. CIDU, the Interdisciplinary Centre for Urban and Regional Development of the Catholic University of Chile was an introduction to the theory and practice of parliamentary socialism in the heady years of the early 1970s before the tragedy of September 1973; this was replaced by the sobering atmosphere of the Social Development Division of the Economic Commission for Latin America in Santiago; in turn, ECLA's lessons were supplemented by the stimulus, challenges and frustrations of directing a UNDP – Office of Technical Co-operation project on national–regional development in Quito, Ecuador.

A decade of interchange at McGill with colleagues and students in the Geography Department and the Centre for Developing-Area Studies, as well as those in the social, physical and agricultural sciences, has added new perspectives to the Third World work. It has also provided a platform and base for wider contacts with NGOs, trade unions, research institutions and universities in Canada and internationally. Of course, the intellectual exchanges are only part of the much broader experience gained from observing, discussing – and very occasionally, participating in – the labour of those in rural and urban environments who, in large part, play the central role in the pages which follow.

Finally, a book such as this puts heavy demands upon the families of the writers. My own family has suffered, from time to time, the slings and arrows of outrageous demands and I take this opportunity to express my gratitude to them, and in particular to Rosalie. To her I offer this book.

1 Third World urbanization: a changing scenario

Introduction

More than thirty years have passed since a pioneer series of surveys of the urbanization process in the Third World first drew attention to the dramatic growth of cities and the significant differences from the experience of the developed countries (see Hauser 1957). At that time it seemed highly improbable that all the Third World would repeat the western experience, classical of a 'true' urban revolution in the sense of a transformation from rural to urban societies; as a consequence one of the authors devised the term 'pseudo-urbanization' to describe the process of urbanization occurring in the Third World (McGee 1967).

The reasons for arguing for this process of pseudo-urbanization were threefold. First, there was the *demographic fact* that the *volume* of population involved in this growth of urbanized societies in the Third World was so much greater than that which had been involved in the western experience. The creation of urbanized societies in the Third World raised real questions concerning the size, form and viability of the larger urban centres if the prevalent patterns of urban primacy persisted. For instance, if the larger Third World urban centres were to experience the same pattern of growth as a London or a Manchester, one might expect to see cities of thirty million plus by the year 2000. At this time the rates of natural increase in the main urban centres remained high and were not falling (largely as a result of medical improvements) while the flow of people from the countryside was often large. Thus the demographic underpinnings of the Third World urbanization posed a fundamental doubt concerning the possibility of the 'true' urban revolution.[1]

Second, it was argued that the economic characteristics of these Third World countries which had been incorporated into an international economic system were very different from those of the developed countries. The manner of this incorporation had created an economic structure (oriented to the export of raw and semi-processed materials and the import of manufactured goods from the developed countries) in which it seemed unlikely that the Third World would experience an 'industrial revolution' similar to that of the developed countries. This reflected very much the early dependency thesis of a stagnating periphery which capitalism could not industrialize.

In the early 1960s there seemed no great advantage for the developed countries to export industrialization.[2] For most Third World countries at that time the failure to create sufficient modern sector employment opportunities left little alternative to the absorption of the growing urban population into the service and small-scale craft sectors. In contrast to the process of tertiarization which emerged in the later phases of western urbanization, that in Third World cities involved the movement of city populations into low-income, low-productivity service sectors, such as small-scale retailing and transport activities. Thus urban involution would accompany rural involution on a path towards increasing immiseration.

Finally, it seemed doubtful in the early 1960s whether most cities of the Third World could provide the institutional frameworks in which new entrepreneurial classes would emerge. The differing historical experiences of Asia and Latin America demanded some modifications to this assertion. In the Asian countries, recently emerged from colonial rule, the concern with nationalism initiated a period in which cities appeared to become 'cult centres'. In such cities the concrete symbols of nationalism were an indication of the diversion of societal energies from economic development. In Latin America this process had begun much earlier, and, by the 1950s, indigenous élites, particularly in the southern cone countries, had been involved in the promotion of collaborative economic development for some decades. It was therefore hardly surprising that when Asian élites were concerned with custodial goals, Latin American planners were formulating programmes of import-substitution development.

Even this depiction is grossly oversimplified and generalized to all parts of the Third World. In a series of subsequent publications both authors have modified this macro picture (see Armstrong and McGee 1968; McGee 1971c, 1979a, 1979b, 1981b; Armstrong 1980; Armstrong and Bradbury 1983).[3]

Changing patterns of Third World urbanization

During the decade of the 1970s there occurred a series of developments which played a major role in the evolution of the world economic system and the patterns and problems of urbanization in the Third World. Of prime importance were the slowdown in the major capitalist economies and the manifold increase in oil prices which contributed to instability in the structure of the world economy. As Logan and Salih comment,

> The simultaneous occurrence of a number of major events in the early 1970s – the move to flexible exchange rates, the sharp increase in oil prices, the commodities boom and the rise in cost of labour in the industrial countries created a set of new conditions. (Logan and Salih 1982, 3)

These new conditions have had important structural and spatial impacts on both the industrialized and Third Worlds.

The developments which have most notably affected the urban system in the Third World are: the technological changes in some industries which are making possible the dispersion of manufacturing to low labour-cost Third World nations; the growing interdependence in the world economy which manifests itself in an increasing transnational control of trade and investment; the growth in volume and movement of capital and finance; and the slowdown in economic growth in the advanced capitalist economies which has coincided with the rapid growth of some Third World countries.

These changes have been facilitated by state initiatives in many Third World countries. The cyclical downturn in the international economy after more than two decades of almost uninterrupted growth throughout the capitalist system has highlighted the extent of such changes. Transnational investment and bank loans in the newly industrialized countries (NICs) have continued apace (especially in the industrializing economies and modernizing agricultural sectors of Asia and Latin America) despite the slowdown in manufacturing output and growing unemployment in the industrial capitalist economies of North America and western Europe. But economic growth in many Third World societies has been bought at the price of a rapidly growing, in some cases apparently uncontrollable, financial indebtedness. Brazil, Mexico, Argentina, South Korea and Venezuela had, by the early 1980s, foreign debts amounting to more than $300 billion. Furthermore, growing developed country protectionism, which limits exports to these markets, has reduced the capacity of Third World countries to meet the steeply rising debt service payments with which they are faced (*Time* 1983, 32–41).

All these changes have had highly irregular spatial and structural impacts, which are revealed in the macrodata of urbanization and structural changes in Latin America and Asia presented in Chapters 4 and 5. The results have led some writers to suggest an increasing convergence in the urbanization patterns of the Third World (see Browning and Roberts 1980) and seemingly to argue for the emergence of urban patterns, relationships and structures which are similar to those of the west. While this may be occurring in major urban centres or central sub-systems (to use de Mattos' (1982) phrase) such as São Paulo, Buenos Aires, Mexico City, Seoul, Taipei, Singapore, Hong Kong and other cities of the NICs, it is difficult to sustain this argument at all urban levels throughout the Third World. Rather, the changes in the international economy seem to be leading to increasing divergence and polarization, between countries and within them. This is reflected in the dynamism and structural characteristics of the urban systems in the Third World.

There appears, nevertheless, to be at least one area in which patterns of convergence can be detected: in the consumer habits, especially of upper income élites but also, increasingly, among middle and lower income groups. Third World metropolises provide the prime locations for imitative lifestyles which find their expression in the adoption of 'modern' consumer habits. Much of this consumption is satisfied by increased imports or the output of transnational branch plants. Consumption of both individual and collective goods reflects the growing convergence on advanced capitalist development styles, and at the same time brings with it a growing dependence on imported products, technology and know-how, as well as, in many cases, a massive indebtedness. Yet the nature of modern urban development has required the adoption of a technology of collective consumption – travel systems, hospitals, schools, etc. – which has created this convergence on the development and growth patterns of the global cities in the industrialized world (see Dunleavy 1980). On the other hand the distortions in this sphere of collective consumption and public welfare create extreme inequalities and problems of unequal access for the urban poor and working class of the Third World cities. Thus welfare that involves capital goods develops apace and individual welfare is left to the family, the community and the informal economy. This process is quite acceptable to the state in many Third World countries for it avoids commitment of social overhead capital. Much of this individual and collective consumption is concentrated in the principal cities, but it is also now being diffused to the populations of regional centres, market towns and even rural areas.

Simply stated, this model of the world economy operates in three interacting spheres of capitalist expansion: production, circulation and consumption. Within the production sphere we note an increasing divergence which is most generally expressed in the term 'the international division of labour'. In the sphere of circulation there are improvements in transportation and in information flows, in television and education for example. In the sphere of consumption there is increasing convergence, which is most obvious in the built environments, transportation and lifestyles. Thus Donald Duck, Coca Cola and Tupperware have become ubiquitous items of consumption throughout the Third World (Dorfman and Mattelart 1975).

It appears that it has become necessary for the capitalist system to extend the techniques of mass marketing beyond the industrialized societies. The growing populations, the rising incomes of many middle and upper income groups and improved literacy and education have provided receptive markets in the Third World for a wide range of aggressively advertised modern sector consumer goods.

It is recognized that there is a class element to this process (see Roncagliolo and Janus 1981, Filgueira 1981). The styles and patterns of consumption of the national bourgeoisie, which imitate those of the advanced capitalist societies, are held up as models which those on lower incomes seek to imitate. In addition, the latter find it increasingly difficult to avoid such consumption since the advance of the capitalist sector is rapidly undermining the capacity of many non-capitalist production systems to offer a viable alternative to the consumer goods of the corporations. Wheat replaces rice and corn as a major staple of Third World cities, and radio, television, cinema and pop music supplant the traditional entertainments of oral history, theatre and folk music.

This pattern of convergence between countries is also repeated within Third World societies. Modern sector production and distribution activities are implanted mainly in the metropolitan centres so that consumption needs in regional cities, towns and rural areas must increasingly be supplied by factory goods either produced by the capitalist sector of the primate city or imported from the industrialized economies.

This process of consumer diffusion is, therefore, a highly effective element in deepening the levels of Third World dependence and indebtedness – socially, regionally and nationally – as Filgueira (1981) has explained. We do not wish at this stage to do more than introduce the convergence-divergence theme and point out that we are discussing another facet of the process of capital accumulation. With the modernization and westernizing of local lifestyles, the creation of new wants at every level, and, above all, the changing of consumer behaviour, the way is opened for capital, both transnational and local, to penetrate Third World markets with increasing intensity. This discussion will be taken up again more extensively in Chapter 3.

It now appears that the complexities of development in the world capitalist system, together with their variable spatial and structural impacts, make the earlier thesis of 'pseudo-urbanization' unacceptable when applied to specific societies. If one considers the Third World as a whole, however, some of the elements of the concept may be valid. Table 1.1 shows that, while the world will become urbanized by AD 2000 to the extent that more than 50 per cent of the world's population will be resident in urban places, this urban population will be most unevenly distributed, with the more developed countries reaching levels of urbanization of 80 per cent and the less developed countries (the Third World)

Table 1.1 Total world population by urban and rural residence, 1920–2000 (in millions and percentage of total population)

	1920		*1940*		*1960*		*1980*		*2000*	
	000,000s	*%*	*000,000s*	*%*	*000,000s*	*%*	*000,000s*	*%*	*000,000s*	*%*
World										
Total	1810	(100.0)	2295.1	(100.0)	2986.0	(100.0)	4374.1	(100.0)	6344.3	(100.0)
Urban	360	(19.9)	570.0	(24.8)	1012.1	(34)	1806.8	(41.3)	3208.0	(50.6)
Rural	1450	(80.1)	1725.1	(75.2)	1973.9	(66)	2567.3	(58.7)	3136.2	(49.4)
More developed countries										
Total	672.7	(100.0)	820.6	(100.0)	975.8	(100.0)	1181.0	(100.0)	1360.2	(100.0)
Urban	260.0	(38.7)	385.0	(47)	572.7	(58.7)	834.4	(70.7)	1092.5	(80.3)
Rural	412.7	(61.3)	435.6	(53)	403.1	(41.3)	346.6	(29.4)	267.7	(19.7)
Less developed countries										
Total	1137.3	(100.0)	1474.5	(100.0)	2010.2	(100.0)	3193.1	(100.0)	4984.1	(100.0)
Urban	100.0	(8.8)	185.0	(12.5)	439.4	(21.9)	972.4	(30.5)	2115.6	(42.4)
Rural	1037.3	(91.2)	1289.5	(87.5)	1570.8	(78.1)	2220.7	(69.5)	2868.5	(57.6)

Source: Hauser and Gardiner (1980).

Table 1.2 Population of cities of size 100,000+, 1920–2000 (in millions and percentage of total population)

	1920		*1940*		*1960*		*1980*		*2000*	
	000,000s	*%*	*000,000s*	*%*	*000,000s*	*%*	*000,000s*	*%*	*000,000s*	*%*
World	169.8	(9.1)	288.8	(12.6)	597.2	(20.0)	1177.6	(26.9)	2245.7	(35.9)
More developed countries	137.6	(20.5)	217.7	(26.5)	360.7	(37.0)	554.1	(46.9)	750.4	(55.2)
Less developed countries	32.2	(2.7)	71.1	(4.8)	236.5	(11.8)	623.5	(19.5)	1495.3	(30.0)
Africa	3.1	(2.2)	7.6	(4.0)	26.9	(9.9)	86.2	(18.7)	249.1	(30.6)
Latin America	8.5	(9.6)	18.0	(13.9)	56.1	(26.0)	157.8	(42.5)	354.9	(57.3)
North America	39.4	(34.7)	55.9	(38.7)	107.8	(54.3)	150.0	(60.5)	197.3	(66.6)
Asia	34.2	(3.3)	74.0	(6.0)	182.9	(11.1)	434.6	(17.3)	970.1	(26.7)
East Asia	21.2	(3.8)	46.4	(7.3)	106.5	(13.5)	229.8	(21.1)	431.2	(31.5)
South Asia	13.0	(2.8)	27.7	(4.5)	76.4	(8.9)	204.8	(14.4)	538.9	(23.8)
Europe	74.0	(22.8)	99.9	(26.4)	161.5	(38.0)	232.1	(47.7)	308.2	(57.1)
Oceania	2.6	(30.6)	4.0	(36.4)	7.0	(44.6)	12.4	(52.8)	18.4	(56.3)
Soviet Union	8.0	(5.2)	30.0	(15.4)	55.0	(25.7)	104.5	(39.0)	147.7	(46.9)

Source: see Table 1.1.

only 42 per cent urbanized. But, even at this level of urbanization, the Third World will account for 60 per cent of the world's urban population, a fact which indicates that there will be very large urban agglomerations in which employment problems will abound.

The main feature emerging from the table, however, is that the increase in urbanization levels will still be insufficient to bring about a reduction of the numbers in the countryside of the less developed world; rural population will actually increase from 1037.3 million in 1920 to 2868.5 million by the year 2000. This is in complete contrast to the pattern that characterizes the advanced capitalist countries, and suggests that the urban safety-valve cannot necessarily be turned on in order to absorb those from rural areas seeking to escape growing rural immizeration following on technological displacement.

This dilemma is strikingly revealed in Figures 1.1 and 1.2 which contrast the considerable differences in wealth, urbanization and patterns of population growth between Asia, Latin America and the developed countries of these regions, in turn reflecting the differing historical impacts of the incorporation of Third World countries into the world capitalist system. Table 1.2, which shows the changes in population in cities of over 100,000 between 1920 and 2000, clearly indicates this discrete pattern, with Latin America beginning to approach the levels of urbanization of the more developed countries, and Asia and Africa still remaining less than one-third urbanized. Most important, however, is the dominance of the Asian population giants, China and India, which make up almost two-thirds of the population of the Third World. It is in these giants that the true urban revolution seems least likely and the challenges of what Friedmann and Douglass (1976) call 'agropolitan development' will have to be taken up.

The second buttress of the pseudo-urbanization thesis, relating to the assertion that Third World cities will not undergo the same structural transformation in the various components of the labour force that characterized the industrialized countries of Europe, no longer seems valid for the following reasons.

First, as Browning and Roberts (1980) have pointed out in their analysis of sectoral changes in the labour force, it is doubtful if the European model was imitated even by North America. Thus it appears that only Europe has experienced the historical sequence by which employment was largest first in agriculture, next in manufacturing and finally in services. In North America and Latin America, employment in this century has always been higher in services than in manufacturing. It is obvious that the 'pseudo-urbanization' argument suffered from Eurocentrism.

Second, the prospect of Third World cities failing to become environments in which sufficient modern sector employment opportunities are generated must now be qualified. It is clear that in population giants such as India the ability of cities to provide labour-absorptive environments is highly doubtful. On the other

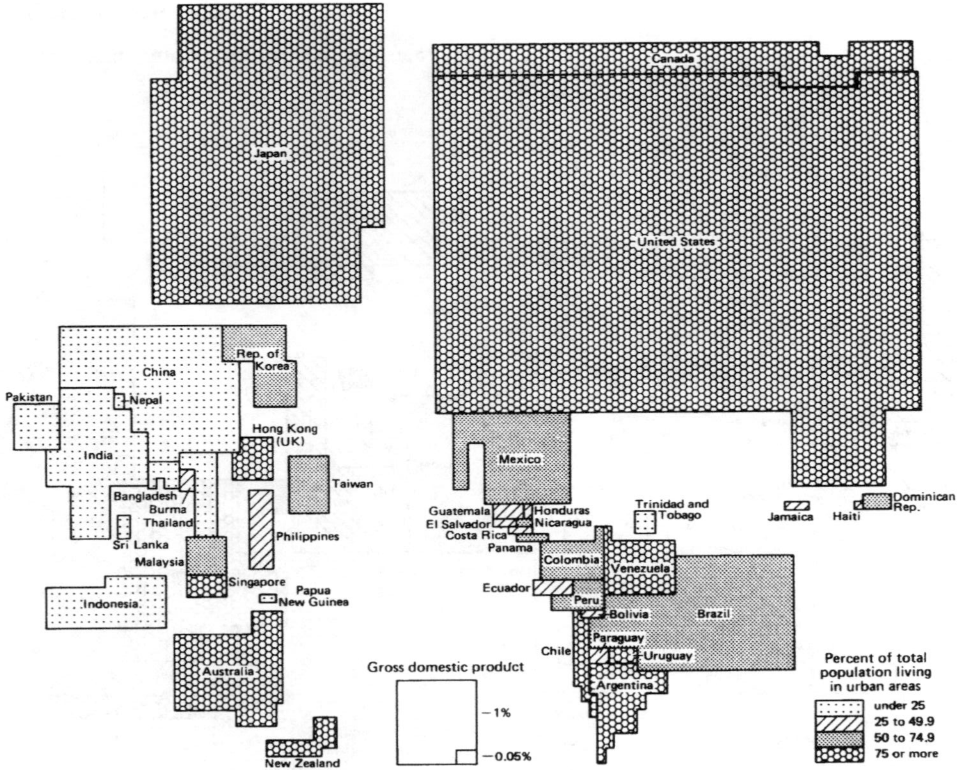

Figure 1.1 Cartogram to show Asia–Latin America: GDP levels of urbanization

Source: World Bank, *World Development Report*, 1984.

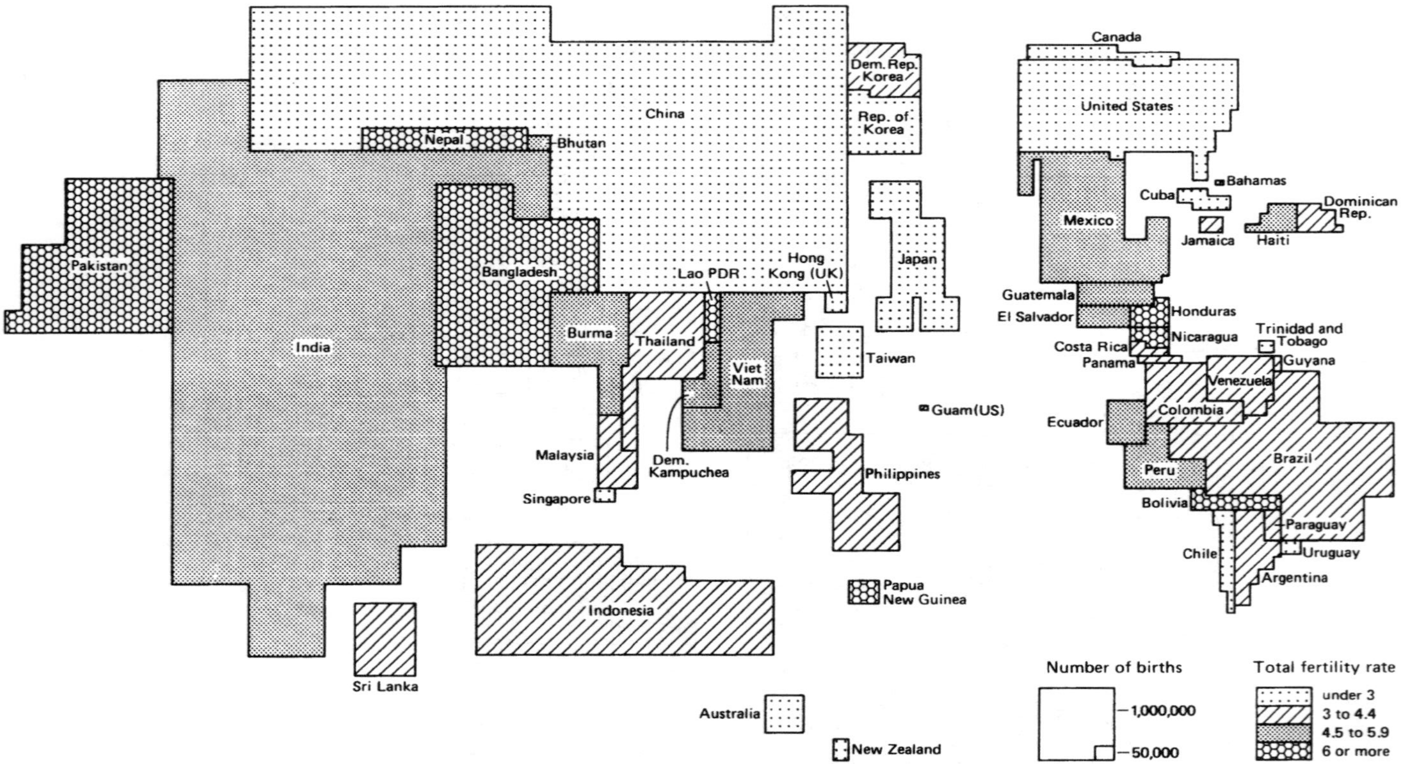

Figure 1.2 Cartogram to show Asia–Latin America: births, fertility rate, 1982.

Source: World Bank, *World Development Report*, 1984.

hand, the rapid growth of modern sector activity in certain Latin American and Asian cities and central sub-systems contradicts the pseudo-urbanization thesis.

Finally, the third prong of the pseudo-urbanization thesis, which relates to the negative role of Third World cities as national 'cult centres' rather than frameworks which aid economic growth, seems ever more inadequate in the face of the changing world economy. Increasingly, governments and the national bourgeoisie are forging alliances with multinational capital which allow the development of indigenous entrepreneurial capacity and the emergence of a technocracy with the motivation to push for economic growth. This is perhaps most clearly shown in the case of Singapore, where a modern postwar capitalist activity has facilitated rapid economic growth, great changes in the built environment of the city and major structural changes in employment.[4] The emergence of these élites does not necessarily mean that social or economic inequality is lessened. Indeed, there is evidence in many cases of the emergence of a new and increasingly aware proletariat which, side by side with the older proto-proletariat, poses political and social challenges for the élite.

New approaches to Third World urbanization

In the next two chapters we attempt to set the theoretical approaches to the study of urbanization within the broader context of development with respect to the investigation of Third World urbanization. The 1970s have been characterized by four new approaches that have greatly enhanced our theoretical understandings of the urbanization process.

First is the emergence of the 'world political economy' approach to the study of urbanization, as represented by the work of Frank (1967), Amin (1974), Wallerstein (1974), Walton (1976a, 1976b), Portes and Walton (1981), Cohen (1981) and many others, which raises questions about the manner in which urban systems and urban centres in Third World countries are a reflection of the role that the national states play in the international economic system. Although the major writers from this perspective have shared an ideological distaste for capitalism, a sharp difference of opinion has emerged between such writers as the late Bill Warren, and Frank and Amin. Frank and Amin focus upon the role of Third World cities as the institutional structures that permit the accumulation of capital in the Third World countries and its siphoning-off to the metropolitan centres of the developed countries. According to this perception, Third World cities play a crucial role in the underdevelopment of the Third World. Warren (1973, 1975), on the other hand, argues that this integration of Third World countries has created the conditions for rapid independent capitalist development and has set in motion a definite process of industrialization that focuses on the major urban places. His arguments have been further advanced in the work of Roberts (1978) on Latin America. The significance of this debate is that it indicates how developments in the international economic system have ramifications on the urban systems of Third World countries, and suggests they be taken into account in research proposals.

One further aspect of this political economy approach is to focus upon the role of international urban hierarchies in the process of developing regional systems. Thus, in considering questions of urban development in the Association of Southeast Asian Nations (ASEAN) region, it is useful to evaluate the role of Singapore as a regional centre. The internationalization of the world economy has increased the importance of such regional centres.

A second major theoretical and empirical contribution to an understanding of the urbanization process in Southeast Asia has been made by demographers, geographers and anthropologists who have begun to look critically at the processes of population movement between rural and urban areas. Research by Hugo (1975), Goldstein (1978) and Pryor (1979) utilizing censuses and survey data have contributed to a much more sophisticated picture of the process of population mobility to and from urban centres. This research has drawn needed attention to the persistence of circulatory migration. It has also provided a careful analysis of urban residence, revealing the weakness of distinction between rural and urban residence. These important contributions should enable a much more sophisticated analysis of the 1980–1 round of census data than would otherwise be possible, which will be of great assistance to policy makers concerned with rates of urban growth and the characteristics of rural-urban migration.

A third theoretical development, which has emerged from a broad body of development theory but has important implications for policy aimed at smaller urban centres, has been the growing concern about the persistent poverty of sizeable proportions of the Third World populations (most of them located in rural areas) and the need for some programme of income redistribution that can deliver basic needs to these deprived populations. Although there has been much debate on the best manner for this to be done and its effect upon economic growth (summarized by Singh 1979), it is widely accepted that an efficient urban system plays a major role in delivering these basic needs (see Sandbrook 1982). At least one writer, Michael Lipton (1977), has argued that most national policies in developing countries are biased in favour of the larger urban areas despite a rhetoric that emphasized the role of smaller urban centres and rural development. Other writers, notably Lo and Salih (1978), have grappled with this problem of urban bias by advocating policies that would combine certain elements of selective regional closure with rural development. These policies would enable small towns to provide more services and, it is hoped, employment, which would lead to the wider provision of basic needs. The policies' major contributions are the emphasis on integrated development and a clear understanding of the function of urban systems in a nation's economy.

The final theoretical development relates to the concept of the informal sector (see McGee 1971c, 1973, 1974, 1976; McGee and Yeung 1977). The idea of the informal sector has emerged from the tradition of dualistic models of the structure of Third World economies put forward by such writers as Boeke (1953) and Lewis (1958). The specific circumstance that gave rise to the growing interest in the informal sector was the concern over the inability of many Third World countries to generate sufficient wage-earning opportunities during the 1960s and the realization that many people were employed in what were essentially family or small-scale activities. These activities involved production (agriculture and manufacturing), distribution (e.g. vending), and construction (squatter housing). Conventional economic attitudes towards these activities have been unfavourable. It was argued that the low incomes and low productivity of the informal sector were a drag upon economic development. But a series of studies during the 1970s, notably by the International Labour Office (1972) and Hart (1973), have questioned this pejorative view. Increased concern with income distribution and meeting basic needs has buttressed arguments that policy makers should adopt policies that would encourage entrepreneurial activity and capital accumulation in the informal sector. Supporters of policies favourable to the informal sector are attempting to persuade policy makers that the persistence of the informal sector is a consequence of a rapidly growing labour force and the wage sector's inability to create sufficient jobs. Some writers (e.g. Squire 1979, Roberts 1978) have argued that many Third World countries have done much better at creating jobs in the wage sector than had been thought possible earlier. For instance, Squire has calculated that between 1960 and 1970 industrial employment in the less developed countries grew at twice the rate experienced by developed countries between 1900 and 1920. But fast growth of the population and the labour force has retarded the transformation of the industrial structure of the less developed countries' labour force (Squire 1979, Table 6).

In the last few years several researchers have criticized the concept of the informal sector. They have argued that for theoretical purposes the concept is too descriptive and leads to a tendency to ignore the relationships between the formal sector and the informal sector. They also argue that it is profitable for the formal sector and the state to allow the persistence of the informal sector. (See Gerry 1974; Bromley and Gerry 1979; and McGee 1978a, 1978b, 1979a, 1982 for a summary of these arguments.) This may seem to be a contradictory position, for they have also argued that the state has adopted policies designed to eliminate activities of the informal sector. In fact, the wide variety and complexity of activities in the informal sector often cause the state to adopt contradictory policies. For instance, the state may carry out clearance operations against cooked-food vendors while enacting price subsidy policies on rice that enable vendors to purchase rice and sell it at low prices.

Other critics have drawn attention to the difficulty of applying the concept of the informal sector to the Third World, arguing that there is persistent confusion over whether it is the unit of economic organization or the form of employment that is being used to define the informal and wage sectors (Mosley 1977). Yet other critics (e.g. Santos 1979) have drawn attention to the lack of dynamism in the dualistic model, suggesting that it would be more accurate to conceptualize the economic structure of Third World cities as different circuits of economic activity that are in a constant state of flux and interaction. While Santos conceptualizes these circuits as a lower and upper circuit, it is not difficult to perceive them as one circuit of activity as, for instance, with transport and retailing.[5]

Finally, the informal/formal sector dualism has been criticized particularly as it applies to the labour markets by writers such as Breman (1976), who argue that it oversimplifies the 'fragmented' labour markets that characterize many of the Third World countries.

These conceptual advances of the 1970s have enhanced our understanding of urbanization processes and as a consequence have implications for both research programmes and policy options. To reiterate, these advances have, first of all, placed increased emphasis on the understanding of national urban systems as being, in part, a reflection of the manner and incorporation of developing countries into the international economic system. This process of incorporation has important effects on the urban hierarchy such as primacy, the functional characteristics of the urban system, and the labour force structure and distribution. Second, the conceptual advances have placed increased emphasis upon the mobility of populations both within rural and urban areas and between them. This new understanding of the mobility of population enables more accurate analysis of the pace of urban transformation, the relative importance of different size centres in the urban hierarchy and the relative importance of rural and urban linkages. Third, the revised concepts of the 1970s have placed increased emphasis upon the study of the processes that create both spatial and structural inequalities in countries, and have led to a consideration of the manner in which the states may intervene in these processes so as to reduce such inequalities. Finally, they have led to increased emphasis upon a critical evaluation of the models that are used to conceptualize the economic structure of Third World cities, and in particular point the need for more detailed studies (e.g. micro-level of economic activity) in order to build more sophisticated models.

Persistent problems of Third World urbanization

These developments in the Third World urbanization do not touch upon any of the more pressing problems that the continuing increase of population raises for the operation of these cities. As Brown (1977) has pointed out, the growth and viability of these cities will rest upon three conditions: first, that food surpluses, produced in these countries or imported, will be sufficient to feed the urban population; second, that cheap energy will be available to service the energy needs of these urban populations, and finally that the city and national governments can afford the investment in, and upkeep of, the physical fabric (for example water supply, sewerage, transportation) of the city.

There are reasons for pessimism about all three. Brown has shown how the growing urbanization of the world has profoundly altered the world food trade pattern, so that, while western Europe was the only major food importing region during the late nineteenth and early twentieth centuries, today many of the major cities of the Third World have come to depend on food imports, of which grain from North America and Australia forms a major component. Brown goes on to suggest that this is partly a consequence of the inability of the Third World countries to produce sufficient food at competitive prices for their growing urban populations. The consequences are grave for, he says:

> Dependence by the world's cities on imported food from a single geographic region in a world of food scarcity may bring with it vulnerability to external political forces and climatic trends that would be risky indeed. Yet the trends to urbanization undoubtedly will lead to greater dependence on North American food. (Brown 1977, 269)

In a sense, these import-dependent cities are one of the more salient expressions of the broader societal changes resulting from the internationalization of capital accumulation. Rapid urbanization is, in part, the consequence of the drastic transformation of the agrarian sector. Agricultural modernization, the swing from basic foodstuffs to export-oriented production, and the expulsion of the peasantry by large-scale commercial producers have revolutionized the rural economies of many Third World countries (George 1976, 1979; Lappé and Collins 1977; Burbach and Flynn 1980; Feder 1976a, 1977).

Self-sufficiency in food production for local consumption is giving way to a growing import dependence. Further, the introduction of mechanized farming methods requiring fossil fuels, together with high inputs of fertilizers and other chemicals, is deepening the energy-dependence of many economies. At the same time, increasing numbers of peasants who once produced for their own needs and for local markets, have become dislodged from the land to swell the numbers in cities reliant upon imported foodstuffs and other basic consumer products. In such circumstances, it seems likely that their conversion to imitative consumer habits is both a reflection of changes in production relationships and induced by consumer behaviour patterns.

Finally, there are many problems relating to the cost of investment in the physical and social fabric of the Third World cities. Many writers have questioned the ability of some Third World countries to afford to run large cities. Indeed, in the Bangkok metropolitan area even the development of a city-wide sewerage system has failed because of the inability to raise sufficient funds. As a consequence, some commentators, particularly those from international agencies such as the ILO, the UNDP and the World Bank, have been encouraging programmes of service provision in Third World cities that place great emphasis upon self-help in housing and services by low income groups. The implications of, and underlying reasons for, these major problems are explained in the following chapters. We argue that many theoretical approaches to the study of the urbanization process in the Third World have been less than adequate because of their failure to examine cities within the context of the operation of a world system in which international capital has been penetrating widely and deeply in response to competitive pressures and the consequent need to establish control over resources and markets.

Before beginning the theoretical section of the book, we think there is a need to make a brief anticipatory comment on our theoretical perspectives. We do this to try to avoid any hasty inclination by readers to label the book ideologically. We had considered the option of emulating the more esoteric studies of some of our colleagues with a lengthy discourse on the present state of the structuralist and post-structuralist (realism) debate. While we do cover some of this ground, we suspect that most readers would prefer to know more about Third World people and social classes at ground level.

Moreover, experience has taught us to treat much of the theoretical abstraction which appears in learned journals with caution; to read it with interest, but also with scepticism. We doubt that the Ecuadorian peasant who uses every resource to keep his family alive and his land intact, or the Philippine street vendor who cannot afford the most basic medical treatment, is deeply concerned with the specific mode of production to which they are attached.

What we attempt to do is focus upon dynamic processes of economic and social change which are sharply focused in the urban centres of the Third World. These changes are the direct consequence of capitalism's cycles of expansion and contraction, over which the peasant, the hawker and the squatter, as well as the wage earner, have little direct control.

2

Development theory and urbanization: rethinking the paradigm

> The theory of underdevelopment and development can only be the theory of the accumulation of capital on a world scale. (Amin 1974, 230)

Prologue

The role of cities in both capital accumulation and the generation of dependence, structural inequality and poverty is part of the larger history of the unequal relations existing within and between societies. It is, for us, a central role, the study of which helps us to understand the great variety and diverse dimensions of inequality. It has not, however, assumed such an important place in many schools of thinking in underdevelopment studies.

Dependency writers, beginning with André Gunder Frank, used the framework of metropolis-satellite relationships at different levels – from the international core of industrial capitalist societies to the rural hinterlands of the Third World. Cardoso and Faletto in their *Dependencia y Desarrollo en América Latina* (1969) took this further in making a valuable contribution to the understanding of the great diversity of forms of regional dependence.

Structuralism, too, has provided revealing perspectives for the investigation of underdevelopment but often at a high level of abstraction. Even with case studies, the theoretical construct appears the more dominant and the varieties of experience, especially in the study of cities, are often lost. The work of Castells is a partial exception here although, again, the diversity of Third World urban situations tends to be overshadowed by the dominant theoretical model.

This chapter is an attempt to review the currents of theory dealing with the questions of inequality and underdevelopment over the past two decades. It serves a double purpose: providing a general framework for the more specific study of cities in the process of capital accumulation and inequality-creation; and making the point that cities deserve far more attention than development theory has hitherto accorded them.

Introduction

Some years ago we wrote a paper under the title of 'Revolutionary change and the Third World city: a theory of urban involution' (Armstrong and McGee 1968). In this paper we argued that the major variable accounting for the pace of revolutionary changes was the extent of capitalist penetration of Third World societies. By capitalist penetration we meant the insertion of modern capital-intensive corporate activity, together with the creation of an urban and a rural working sector of wage earners alongside an existing sector of non-wage earners.

Of course, we were drastically oversimplifying capitalist expansion in portraying it primarily in terms of the growth of a proletariat. A more accurate model would have also attempted to show the increasing concentration and changing characteristics of capital formation and ownership. However, by reference to a comparison of Cuba and Indonesia, we argued that Third World societies in which a large wage-earning sector had been created and in which class relations had thus been polarized were more susceptible to revolutionary change than those in which the growth of the wage-earning sector had not been so extensive.

In the case of Cuba, we pointed out that the expansion of capitalist production had largely eliminated the pre-capitalist sector and created a class of rural wage earners, and an urban proletariat and proto-proletariat which supported – or at least were neutral towards – the revolution. In Indonesia, the process of direct capitalist penetration of city and countryside was much less advanced. The persistence of what we then called 'bazaar' and 'peasant' systems in both rural and urban areas militated against revolutionary change. Thus the failure of the 1965 'Gestapu' coup.

In explaining this process further we employed a model of metropolis-periphery relationships developed by Andre Gunder Frank (1967), in which the Third World city system was regarded as the crucial link in an accumulative and disseminating chain of capitalist relations stretching from the international metropolises to the countryside of the Third World. The second major concern of the paper was with structural contradictions of this capitalist penetration. The rapid growth of the majority of Third World primate cities (often at rates more than double those of natural population increase) and the failure of the modernization process in many Third World countries to create sufficient employment opportunities in the wage-earning sector were creating a substantial class of low productivity petty producers and distributors.

We felt that the capitalist modernization of agriculture would lead to the acceleration of rural-urban migration and further expansion of the petty producers and distributor groups in the Third World cities. But we were not prepared to predict that, in most cases, the consequences of this process would be revolutionary, except in the long term. Rather, we borrowed from Geertz's concept of rural involution in the Javanese countryside and extended it to the capacity of the urban self-employed to practise labour-absorptive activities. Nevertheless, we felt that his 'bazaar sector' would eventually have to cease involuting. Then the contradictions arising from capitalist penetration would come to a head and some form of significant structural change in the society would have to occur.

The developing development paradigm

In the period since that paper was written, the issues it raised have become central to many concerns in development studies.[1] In the late 1960s a growing disillusionment with the conventional models of developmental change occurred. The failure of the 'Sixties Development Decade' underscored the fact that the linear Rostowian stage models of repetitive economic development for the Third World were false and that the theories of westernized imitative modernization were suspect. It had also become evident that the mix of elements of the urbanization process in the Third World was very different from the imitative stylizations of orthodox development models (see McGee 1971b, 1976 for a review of these factors). The persistence of stagnation in traditional agriculture, the structural pressures faced by small producers confronted by the penetration of agribusiness,[2] and the continued extension of low productivity activities – in a nutshell the persistence and, indeed, constant growth of poverty in the Third World[3] – led to a challenging of the conventional developmental wisdom.[4]

This challenge took two forms. The first manifested itself in research and policy prescriptions emerging from international agencies such as the International Labour Organization, which sought to focus on the low productivity sectors of Third World countries by resorting to the consciousness-raising device of the 'employment crisis of the Third World'. The structural problems of labour force absorption in Third World cities which we had focused upon in our 1968 paper were identified in this work; and a dualistic model of the economy of these countries distinguishing between a 'formal' and 'informal' (modern-bazaar-peasant) sector was utilized. (See ILO 1972, Hart 1973.)

Despite its low productivity, the informal sector, which the older development models had seen as a blockage to economic growth, was now perceived as dynamic. Policy prescriptions which attempted to encourage government retention of the sector and redistribution of public sector investment towards it were promoted. Later, other international agencies climbed onto the bandwagon. To the 'employment crisis' were added a 'poverty crisis', an 'energy crisis', a 'shelter crisis', a 'food crisis' and, recently, an 'urban crisis'.[5] These suddenly conceived crises were rarely seen as part of a larger process of systematically created inequality.

The second form of challenge was the surfacing, with some degree of respectability, of an approach to development problems which stemmed from dependency theory. As the theorist and planner Mahbub Ul Haq has observed (1976, Chapter 1, *passim*), development fashions have succeeded each other in rapid turnover since the Second World War as theorists and practitioners sought the correct formula for successful development strategies. In a bewildering variety of analyses and interpretations the one constant appears to be the assumption that the world is divided into two clearly separate parts. The titles given to this bifurcation have evolved from 'advanced-backward' to 'industrialized-primary producing', 'traditional-modern' and 'developed-underdeveloped' (or, euphemistically, 'developing'); but always the underlying postulate is that all societies are contained within either the advanced industrial core or the lagging periphery.

Such an approach clearly has certain implications for policy making. Much of the analysis in the various conventional development streams theory has focused upon strategies which offer the underdeveloped periphery the means to enter into the modern world. Essentially, it is assumed that the underdeveloped countries will follow the same path as the nations of the centre. Urbanization, industrialization (either through policies of balanced or unbalanced sectoral and regional growth), demographic policies aiming at population control, agricultural change and modernization, open market policies offering incentives to foreign capital – all have as their stated objective the narrowing of the gap which divides the two worlds. Such assumptions find their way into the planning and development policies of the United Nations, regional agencies, the World Bank and private development institutions operating in the Third World. In recent years, with the inadequacy of such approaches becoming evident in practice, some attempts have been made to refine the concept to take account of the complexities and different levels of national development. A Fourth World has been added to the Third, and, indeed, some writers have attempted to delineate up to seven worlds.[6]

ECLA, the dependency challenge and underdevelopment theory

Before developing these themes in greater detail, we need to retrace some of the steps leading to the alternative approaches to unequal development. One of the stronger and more persistent challenges to orthodox development and international trade theory arose in the periphery itself. From the late 1940s on, some of the economists of the Economic Commission for Latin America (ECLA) began to question much of conventional development theory formulated in Europe and North America. Their questioning of the assumptions of free trade theory (and, more recently, of Chicago school monetarism) and their alternative proposals to restructure Latin America's economies by promoting internalized industrial development presented a clear challenge to those orthodox development styles based on comparative advantage and international specialization of function among nations (Rodriguez 1979).

Above all, the ECLA economists formalized the concept of a world-system divided into 'centre' and 'periphery' and gave it explicit status. A succinct summary of this approach appears in the work of Pinto and Kñakal who define the principal difference between economies of the centre and the periphery; this, they argue, lies in the nature of the development dynamic which arises as a consequence of internal forces within the core societies but not the peripheral societies. Consequently, the developed societies are able to exercise significant influence over development in the periphery (Pinto and Kñakal 1972, 10). Furthermore, while the 'engine of growth' in the developed nations is endogenous and allows for a high level of independence of international economic pressures, in the periphery growth derives from the export sector, thus leaving those societies vulnerable to international trade and capital movements.

Up to this point, the centre/periphery concept appears to reside in the realm of international geo-economic relationships. However, Pinto and Kñakal give it a rather different character by arguing that the term is not completely synonomous with the 'developed' and 'developing' contrast. Although all central economies are developed, not all developed economies are central. Among the latter are to be found the 'developed periphery', the old colonies of Britain – Canada, Australia and New Zealand. They are not part of the centre and suffer certain dependence in comparison with the central societies, but structurally they are developed and can count on their own internal dynamic to promote national development (Pinto and Kñakal 1972, 9–10).

These three countries are defined as developed by virtue of their high incomes, the diversification of their production structures, and, in some cases, their diversified external sectors. Such societies are, therefore, clearly part of a 'developed periphery' although the authors admit that the term does not really satisfy the complicated situation found, for instance, in Canada (Pinto and Kñakal 1972, 68).

This is a useful synthesis, yet doubts must arise as to the value of the 'centre/periphery' concept in analysing the complexities of the world economy. At one level it is used to account for the structure of international economic relationships. At another it enters into a discussion of the location of countries with their different levels of development. It might be asked that if Australia and New Zealand (not to mention Canada) are 'developed peripheries' because of their geographic distance, where is Japan in all this?

Other writers have pointed out the terminological and conceptual difficulties which face those who have used a centre/periphery model to criticize orthodox development theory in recent years, as a number of writers have pointed out (see O'Brien 1975). Does the term 'periphery' define a certain group of nations in the world economy in terms of their exchange relations with others at the 'centre'? Or does it, on the other hand, refer more to their internal structures of production and the social relationships associated with them? Or is its connotation essentially spatial?

Moreover, the dividing of the world into two polarized parts suggests a clear split between two homogeneous blocs. The implications are that the members of the periphery are frozen into a situation where the 'development of underdevelopment' leaves little alternative for internal capitalist evolution, and the only outcome for the exploited societies is shattering socialist revolution. This may well be an ultimate consequence in some cases, but in the meantime there are many other phases associated with the internationalization of capital which are neither so linear nor predictable; the recent experiences in Brazil, Argentina and Mexico and the cases of Taiwan, South Korea and Singapore all provide instances of the variability of capitalist growth in the Third World.

A major difficulty, as with much writing of the dependency school, is that the argument remains at the level of exchange relationships and the problems created by commercial inequalities among nations. The focus is directed to unequal trading links and the consequent transfer of economic surplus (Baran 1957) or value (Emmanuel 1972) from the peripheral nations to the centre, rather than to the internal social, economic and political structures and relationships. From there, it can be a relatively simple progression to the conclusion that the principal contradiction in the functioning of the capitalist system has passed from class confrontation to conflict between nations. Emmanuel writes, 'My subject is the "exploitation" of one nation by another, not the exploitation of man by man' (Emmanuel 1972, 303).

Samir Amin's early work (1974, 1976), takes the analysis beyond this, and is critical of Emmanuel's emphasis on inter-societal relationships as the primary explanation for uneven development. Nevertheless, the main themes of Amin's argument are developed within the framework of unequal relationships between the dominant and dependent nations. The debate is sustained largely at the level of international exchange relationships while questions of internal production structures and the evolution of social relations play a subordinate role in the discussion.

The conclusion that the centre/periphery contradiction is the crucial one for contemporary capitalism gives rise to a number of secondary implications. One is that class conflict in the societies of the centre has been subdued by the very success of capitalism's exploitative accumulation on a world scale. As a result, the working classes of the wealthy nations have been bought off at the expense of the poor nations and have little reason to change such a comfortable situation. To argue in this way is to underestimate the long history of class struggle and confrontation that has characterized the evolution of the industrial working class. Any gains made have been extracted, for the most part, by trade union and political action, rarely by benign concessions from above.

A further difficulty arising from the counterposing of the two blocs is that unlikely bedfellows appear in both. New Zealand and the United States form part of the undifferentiated centre, on the assumption that they are both independent and autocentric societies. Argentina and Ethiopia are both categorized as members of the dependent, underdeveloped capitalist periphery.

One further question needs to be asked. If the obstacles to development created by unequal exchange and the expatriation of economic surplus from one nation to another were resolved through some new international order, to what extent would this reduce inequalities between classes within peripheral societies? An improvement in the terms of trade for Third World raw materials and foodstuffs *should* ameliorate domestic social conditions.[7] However, given the structure of ownership of the means of production in agriculture, commerce, finance and industry, and the distribution of economic and political power in most Third World societies, there is no guarantee that the benefits of larger export receipts would trickle down very far, or in great volume. Examples from the work of Griffin and Khan (1977) in Asian and Latin America societies indicate that there is no constant correlation between increased economic growth rates and a more equitable distribution of the fruits of that growth. Indeed economic expansion may exacerbate inequalities in a society.

The difficulty with emphasizing unequal exchange at the expense of other relationships is that it treats only one part of the total process of capital accumulation. Given the internationalization not only of commercial capital, but also of finance and production capital (Palloix 1977), the unequal exchange argument seems inadequate as a measure of the causes and processes of uneven development. As Radice points out,

> the appropriation and redistribution of surplus value cannot be understood solely in terms of the circulation of commodities (trade), nor of the circulation of money (financial flows), nor of the production process (international division of labour); the accumulation of capital involves *all* these.
>
> (Radice 1975, 17)

The need for a more flexible analysis of the internationalization of capital and its uneven impact on different societies is evident. Wallerstein discusses the emergence of a category of countries within his world-system which form

> *intermediate* elements in the surplus-extraction chain which gain at the expense of those at the *core* of the system. In present-day terms, this means among other things a shift in the relative profit advantage to the *semi-peripheral* nations. (Wallerstein 1976, 464)

This suggests that a bipolarized framework fails to uncover the diversity of world capitalism.

Nevertheless, because it, too, concentrates on external relationships among countries rather than analysing systems of production and their social relationships, the concept of the semi-periphery must try to represent a disparate and unwieldly collection of thirty nations ranging from Canada to Algeria, China to Chile. Wallerstein, then, refines or adapts the centre/periphery argument but the perspective remains essentially the same. Internal social, cultural, political and economic characteristics of each country play a minimal role, while major emphasis is placed on the position of each society in the international economy.

Dependence as an internal practice

Dos Santos (1973b) takes the concept of uneven and dependent capitalist development a significant step further. While working within a general dependency framework, he is careful to point out that national dependence has much more to do with the social structures and class relationships *within* societies than with the external exchanges among them. 'Dependence', he argues, 'is not the "external factor" which it is often believed to be.' Each situation needs to be examined specifically to determine the dialectic between international relations and national forces.

> It would be too easy to replace internal dynamics by external dynamics. Were it possible, we would be spared the study of the dialectic of each movement of a global process and could instead substitute for the analysis of different concrete realities a generalized and abstract formula. (dos Santos 1973b, 73)

He also cites the Peruvian writer Anibal Quijano, who, in a 1967 ECLA paper, notes that dependence 'is not an external point of reference but rather a fundamental element in the interpretation of our history' (dos Santos 1973b, 73). The concept cannot be understood then only, or even primarily, as an external phenomenon concerned with the dealings between nations. Dependence arises from the interaction between external relationships and the internal social, cultural, economic and political situations of each nation. Without examination and understanding of the internal elements, the main consequence of concentrating upon the phenomena of external relationships will be to leave any conclusions at a generalized level of abstraction.

This is very close to the argument of André Gorz:

> Colonialism is not an *external* practice of monopoly capitalism. It is from the first an *internal* practice. Its victims are not just the exploited, oppressed, dismembered nations, but the populations living in the metropoles, in the dominant countries. (Gorz 1975, 17)

He questions the too-facile division of the world into two camps – the capitalist nations and the 'proletarian' nations – making up two monolithic blocs. Instead,

> the frontiers between development and underdevelopment, between dominant economic powers and dominated populations, between colonizers and colonized, do not pass simply *between nations*; . . . these frontiers exist also *within* each nation of the capitalist world. (Gorz 1975, 18)

Charles Bettelheim, too, has warned against the use of the concept of the exploitation of one nation by another rather than a social analysis of exploitation; the latter expresses a production relationship and so 'it necessarily relates to class relations (and a relation between "countries" is not and cannot be a relation between classes)' (Bettelheim 1972, 301). Poulantzas also argues against the premiss of exploitation of proletarian by capitalist nations, given that 'the relationship of exploitation is that between the bourgeoisie as a whole and the working class and popular masses' (Poulantzas 1975, 151). And in a practical application of such approaches, Marshall Wolfe, in his *El Desarrollo Esquivo*, points out that:

> In fact, the spokesmen of the majority of 'poor' countries have no personal reasons to envy the income and lifestyles of their counterparts in the 'rich' countries, and this, as the press commentaries in the latter societies demonstrate, weakens the credibility of their calls to establish a new international economic order. The same international reports which personify the poor countries and attribute to their leaders the decision to eliminate poverty show that, in general, increases in the wealth of such countries and in the operating capacity of their governments have no positive effects on the poverty of the masses, and that there are other priorities for the distribution of the additional

> wealth and the exercise of power. In reality, the predominant development style and the new international division of labour appear to require, even more than earlier centre-periphery relationships, that the highly unequal distribution of income in the 'poor' countries should be perpetuated.
>
> (Wolfe 1976, 14–15, translated by the authors)

Applying a similar approach to a number of concrete situations, Cardoso and Faletto (1969) examine the implications for the different varieties of Latin American societies. Dependence, they observe, follows from a combination of external and internal elements. Expansion from the industrialized centre encounters a range of national situations at the periphery and this results in distinct amalgams of alliances, resistances, and tensions (Cardoso and Faletto 1969, 34).

As the causes of dependence are in large part rooted in internal social structures and relationships, the dynamic of the dependent economy is strongly conditioned by the systems of alliances among its social classes. On this basis, Cardoso and Faletto separate Latin American societies into two types according to internal evolution and experience within the world capitalist system in the past 100 years.

In this way a distinction can be drawn between societies of settlement and societies of exploitation. In the former, Argentina, Uruguay, Chile and southern Brazil, the degree of dependence upon external elements was lessened by the fact that local ruling classes were able to maintain an important measure of control over their export production sector, as well as their financial, commercial and even industrial activities. Such control was founded both upon a system of internal class alliances and external collaborative relationships with international capital (see Cardoso and Faletto 1969, 40–1). In Asia, Japan is clearly the most successful of this genre, although more recently South Korea, Taiwan, Hong Kong, Malaysia and Singapore have assumed similar characteristics.

Societies of exploitation are, in contrast, characterized by the wide extent of external control over the export enclaves. Much of Asia exemplifies this form of development, while in Latin America the most obvious examples are to be found in the raw material and foodstuff producing countries of Central America, in the Caribbean and in the mineral-exporters such as Bolivia.

Cardoso and Faletto have, then, argued that the societies of Latin America do not comprise a monolithic bloc of underdevelopment, and they are joined in this by writers such as Furtado (1970) and Bambirra (1973), while Warren (1975) argues a similar case for Asia.

An awareness of the limitations of the centre/periphery model is also important for political strategy. If the principal obstacle to a country's development is considered to be its relationship with other societies, it is tempting to assume that its resolution can be achieved by a national unity of people against foreign exploitation. Experience throughout the Third World has shown that such expectations will be illusory in the absence of a social transformation within the society previous to, or concomitantly with, the achievement of political sovereignty. While nationalism may provide a useful vehicle in the achievement of social change through resistance to external exploitation, it is not a substitute for such change.

In the later case studies in this book we will observe a particular chemistry of relationships between internal and external forces which is highly specific to each society. Certainly, this is one of the more complex questions which has arisen in social science, for in most cases it is difficult to distinguish between cause and effect in internal and external events.

So that the complexity of this process can be better understood, the way in which capitalism advances and interacts with precapitalist systems requires further study. While each society has its own specific experience of this expansion, the process has comparable features whether historically in Europe or currently in today's Third World societies. For this reason the structuralists' debate on the evolution of capitalism over the past four centuries in Europe and the examination by the 'modes of production' writers of the changes taking place in the Third World societies of the twentieth century present a continuum which is highly relevant to the study of internal and external elements in the development process.

Class relations and modes of production

Through the 1970s the focus of attention in development studies began to shift from the analysis of external factors to that of the internal structures and interactions which condition societal development. This, as we have seen, had already started with the later dependency writers. Now other voices emerged, concerned to correct what they considered an unbalanced external emphasis in the explanation of Third World underdevelopment (Laclau 1971; Brenner 1977; Taylor 1979; Palma 1978, or, at a more abstract level, Althusser and Balibar 1970; Poulantzas 1975; Anderson 1974, 1980). These new groups sought to return to basic Marxism, through analysis of distinct modes of production and relationships among classes within societies rather than studying circulation patterns and relations among nations. The debates on internal and external forces in social change that had taken place between Dobb and Sweezy in the 1950s were thus rekindled in the 1970s (Hilton 1976–8).

These exchanges have often cast more heat than light and the prose has been heavy and convoluted. But, emphasizing the inherent tendencies – or logic – of capital, of class structures and modes of production and (to use the terminology) their articulation within different social formations, structuralism has reminded us of the nature and causes of Third World underdevelopment.

A writer like Castells can apply this methodology in explaining his own more specific concept of 'dependent urbanization' in terms of relationships between societies as well as within them:

> A society is dependent when the articulation of its social structure, at the economic, political and ideological level, expresses asymmetrical relations with another social formation that occupies, in relation to the first, a situation of power. (Castells 1977, 44)

Castells' purpose is to link the patterns of Third World urbanization with the uneven nature of capitalist development, and, in particular, following Poulantzas, to show that dependence is an expression of internal organization and the articulation of productive and class relationships. The results can be seen in the spatial and socio-economic fragmentation of Third World societies, the macrocephalic character of urban growth, disparities between urban and industrial (employment) development, and the expansion of low productivity sectors. In structuralist language, this is experienced as follows:

> Urbanization in Latin America is . . . the manifestation, at the level of socio-spatial relations, of the accentuation of the social contradictions inherent in its mode of development – a development determined by a specific dependence within the monopolistic capitalist systems. (Castells 1977, 63)

To see the application of these ideas at a more general level, we need to consider some of the definitions put forward by structuralists in recent years.

First, the concept 'mode of production' has a significance which transcends the merely economic elements of the productive system. Its application embodies the social relationships between classes involved, on the one hand, in the production of wealth, and, on the other hand, in the appropriation and accumulation of capital. It also incorporates elements of the superstructure such as the political, legal and administrative institutions, together with the society's prevailing ideology. The last is of importance, as we shall see when we discuss not only the impact of capitalist production on Third World societies but also that of imported values such as the ethic of consumerism.

In addition to defining the mode of production as a relationship between social classes, a further distinction is made between the concept of 'mode of production' and that of 'social formation'. The latter is more concerned with the national units within which coexist distinct systems of production of unequal influence and dynamism and interacting in relationships of domination/dependence. In the contemporary world system the dominant mode of production, capitalism, has penetrated into most social formations of both core and periphery, imposing itself in a variety of hegemonic relationships with non-capitalist modes. The debates revolving around these relationships, or articulations, have been part of the mainstream development critiques over the past decade.

While some of the discussion has become rigid, even obscurantist in recent years, the investigation has certainly shown how these unequal relationships are exploitative of the non-capitalist systems; structuralist analysis has demonstrated the multiple ways in which capitalism is able to use inequalities to expand its control and increase its competitive capacity for the extraction of surplus value. This is helpful even if, as Munck (1981, 172) points out, the logic of the argument is astray to the extent that precapitalist systems 'cannot really be *modes* of production if their conditions of existence are so inextricably linked up with the expansion of capitalism'. What we find rather is a subordinate system with its fragments of the precapitalist superstructure eroded by expanding capitalism.

Applying the concepts

In 1968, we conceptualized Third World societies in terms of a linkage between two coexisting systems, capitalism and non-capitalism with the former dominating the latter. The major distinguishing feature of these two systems seemed to be the type of employment – wage labour versus self-employment. But this approach presents real theoretical and empirical problems; as an explanation it now seems to miss the many subtleties and variations in the interactions between the two systems. It is clear that non-capitalist systems of production are deeply influenced and shaped by the dominant capitalist system but this infiltration has not always led to a change in patterns of employment or labour use. For instance, peasant producers of coffee for the international market are undoubtedly involved in the capitalist system of exchange even though their production may be non-capitalist and they retain, formally at least, their status as independent producers. But, as this involvement in an exchange system under the control of commercial enterprises crucially affects the livelihood of the peasant producers, it is not unreasonable to see them as a subaltern part of the capitalist mode of production.

The contradiction between the formal peasant ownership of a piece of land and the reality of his incorporation into the system of capitalist production and distribution is especially clear under conditions of sub-contracting in agribusiness. The production becomes just one phase in a sequence of activities organized by the corporation to produce, process, store, transport and, finally, sell the goods to the North American, Japanese or European consumer. The degree of exploitation of the non-capitalist producer is not less real for his/her retaining a sense of proprietary independence (George 1976, 1979; Feder 1976b, 1977).

The problem of the 'sustainability' of rural non-proletarian activities in the face of capitalist penetration has become a central issue among researchers in rural areas of Asia. Simply put (as we understand it) the crux of the problem as developed earlier in Chayanov (1966) and Franklin (1969) is that the peasants have been subordinated to capitalism through the reconstruction of their production process to include the production of commodities for exchange value. But their means of production has not been expropriated and their 'moral order' remains essentially pre-proletarian (see Scott 1976). They are thus able to resist the efforts at further penetration of the state and capital (see Shanin 1971). It is argued that traditional systems of labour relations (reciprocity), other social relationships and ideology act to reinforce this resilience. With this argument we have no quarrel for it is patently the case, but it takes insufficient account of the various ways by which the capitalism and the state begin to dislocate the peasantry.

On the basis of a fieldwork of high quality, De Koninck[8] has shown, for example, that the peasants of the MUDA project in Kedah, Malaysia, while experiencing increases in productivity and income, are increasingly losing control of the tools of production, but not their land. The technological changes of the Green Revolution, as a result of which peasants have increasingly lost control of the irrigation systems, the technological inputs and various processes of planting and harvesting, are becoming more mechanized and controlled by managers and entrepreneurs who inform, but are not informed by, the peasant. The consequences are that collective labour is becoming less important, and more off-farm employment is sought. It is first and foremost the labour of women that is displaced by machines and offered at low prices on the market, but men, too, are moving into proletarian employment.

De Konick (1981) argues that this process of losing control of the tools of production is as effective as losing control of land. Thus capitalist penetration at many levels can dislocate non-proletarian work in the countryside.[9] When this is associated with the penetration of consumerism, mere possession of land is no guarantee of the continuation of the peasantry.

Given these considerations, we have adopted a conceptual model in this study which does not rely upon a 'bi-modal' division. Rather, we see each society as a part of global capitalism, but a unique part which has its own particular mix of production and exchange. The internal relationships generated exhibit immense diversity, as will become evident in the case studies later in the book.

Seen globally, the expansion of corporate monopoly capital and neo-colonial relationships in the post-Second World War years has engendered structures of dominant, intermediate, and dominated classes in the Third World societies. Local ruling class alliances have usually evolved from an uneasy amalgam of the traditional and modern groups, including the armed forces élite, the landed oligarchy, indigenous merchant and industrial entrepreneurs, and university-educated technocrats, all of whom control political power and provide the necessary framework to support the process of capital accumulation. In turn, the state and private sector firms are staffed by an administrative and technical middle class with limited decision-making power but which, nevertheless, sees its interests clearly bound up with those of the hegemonic collaborative ruling alliance.

Petras (1975, 296) emphasizes the significance of the national 'intermediaries' who stand between imperial capitalism and the labour force of Third World states; it is a social stratum drawn from the 'propertyless' groups which has both the desire for personal ownership through association with metropolitan or public enterprises, and access to the powers of state.

In a more detailed analysis of the state, Roxborough (1979) suggests that the functioning of the class state can be sought 'in the operation of a set of institutions which socialize state functionaries into the world-view of the dominant class' or, where there are differences of viewpoints, in the 'identity of interests between the functionaries and the dominant class' (1979, 121) which reflects the alliance between a social class and the political élite. Where there are internal divisions that cannot be resolved within the dominant class, however, the way may be opened for the military to 'occupy the state apparatus and rule on behalf of the dominant class as a whole' (1979, 122).

These social classes can be represented by a pyramid. At the base of the pyramid we can distinguish the large but fragmented and politically powerless majority: urban and rural workers, peasants and petty or small-scale producers. The 'intermediate' class is made up of government servants, small indigenous businessmen and the 'technocrats'. At the peak of the pyramid are the political, economic and military élite.

Having generalized this way, we must also stress that there are considerable differences in the degree of social and political organization which result in diverse patterns of alliance and confrontation varying from one society to another. In setting out our present schema of Third World class structures we have argued that, over the long term, capitalism tends to create broadly similar class structures and relationships. This, too, of course, is a generalization which needs to be modified by reference to the diversity of historical experiences and levels of material and social development among societies.[10] It can be refined by reference to the degree of capitalist accumulation, the relationships between the capitalist and non-capitalist sectors and the characteristics of the social classes – all of which will vary from one Third World society to the next (Poulantzas 1974, 148; Roxborough 1979, 76ff.).

As Roxborough notes in illustration of this, in many societies, social and economic inequality, externally created economic instability and low absolute output levels mean that 'it is difficult for dominant classes to develop a great deal of legitimacy. They simply cannot deliver the goods and thereby win the acceptance of the subordinate classes' (Roxborough 1979, 124). Under such circumstances, the long-term tendency towards convergence on similar capitalist structures may be very long indeed, and, in any case will vary widely from society to society. We have therefore to draw a distinction not only between the development trajectories of nations as different, for example, as Argentina and Indonesia, but also among such close neighbours as the societies of Central America, where specific historical experience has given each its own social, economic, cultural, and political structures and relationships.

Returning to our global argument, we see the importance of relating the three-tiered class structure to the patterns of capital accumulation in each society. It is the interaction between the internal situation and the processes of capital penetration from outside which, as we make explicit later in this chapter, gives each society its own unique trajectory of national, sectoral and urban development.

It can be expected that where capitalist production has become virtually exclusive, the predominant form of accumulation will derive from the surplus extracted from labour in the production process. A secondary form of accumulation will be through the process of unequal exchange. Rent will form a third and subsidiary means. However, in societies in which the capitalist system is dominant but not exclusive, the extraction of surplus value through the production process will be less important, while the mechanisms of unequal exchange and rent will play a correspondingly larger role in accumulation.

So diverse are the variations that there will be no two societies with identical patterns of capital accumulation. Indeed, the interaction with the global expansion of capitalism of particular elements of historical experience, political institutions, physical resources, economic structures and socio-cultural relationships within each society will result in highly specific chemistries of development for each society.

These variations in capital accumulation have important consequences for political action. As Petras points out, Third World societies have at least three strategies, based on different class alliances, to deal with capital accumulation (Petras 1975, 297). First, the local ruling class can ally itself with international capital to increase the extraction of surplus through unequal exchange and direct appropriation from production. Second, the dominant groups can siphon off surplus value and, by limiting the share of income going to foreign firms, divert part of the surplus into the hands of the state and the national bourgeoisie. A third possibility is for local ruling groups and the government to ally with the working class and extend control over some areas of production (see Poulantzas 1974, 165–6).

In writing of the corporate monopolization of food commodity production, processing and marketing by transnational corporations (TNCs) in conjunction with internal collaborative oligarchies (ICOs), Clairmonte and Cavanagh (1982, 38–9) leave the impression that strategy 1 is dominant for the food export sectors of the Philippines, Thailand, South Korea and Central America. It is a partnership reinforced by the support of the governments of the industrialized world and by international agencies. But it is also a situation threatened by growing political instability as popular movements arise to challenge the ICO-TNC relationship.

It is evident, too, that strategies may change radically in response to changing class control of political power. For instance, the 1964 coup in Brazil under the aegis of the military speeded the process by which a modernizing bourgeoisie could establish a bureaucratic authoritarian regime to replace both the populist national groups and the older comprador export-oriented oligarchy as the ruling alliance. In another example of the process, the nationalist coalition of the Sukarno era in Indonesia was replaced in 1965 by a grouping of classes that encouraged the free flow of foreign investment. The same pattern also occurred with the imposition of authoritarian capitalist regimes in Argentina, Chile and Uruguay in recent years, although Argentina has reverted again with unpredictable consequences. Obviously, it is also possible for any society to have a combination of strategies 1 and 2 and even 3. Thus Brazil now appears to combine 1 and 2 and Tanzania elements of 2 and 3.

From the point of view of the urban systems that develop under these strategies, it is probable that strategy 1, a heavily export-oriented one, would lead, in certain circumstances, to the growth of a concentrated and polarized pattern of urban places designed, above all, to facilitate the flow of surplus to the advanced capitalist societies. It would typically consist of a relatively large, though probably not highly industrialized, primate city linked directly to extractive zones of agriculture and mineral production. Tacked on to this would be a network of much smaller and (for the dominant mode) economically less significant administrative and market centres. This pattern of urban centres, ubiquitous during the colonial phase of many parts of the Third World, has been described in detail by McGee (1967).

However, with the growth since 1945 of 'dependent industrialization' in the Third World (itself a consequence of the internationalization of direct production capital) new elements have been added to these urban systems. At an international level the growth of qualitatively different cities such as Hong Kong and Singapore provides important evidence of these trends. Within Third World societies the expansion of urban places has led to the formation of complex industrialized regions around the metropolises of Mexico City (Mexico), São Paulo (Brazil), Pusan (South Korea) and Kaohsiung (Taiwan), while Buenos Aires has continued to expand from an industrial base built in the early part of this century. Although the incidence of these industrial complexes throughout the Third World is very uneven, it is a major trend affecting urban systems (Roberts 1978, 74ff.). Before leaving strategy 1, we should note that not all export enclaves have given rise to a concentrated pattern of urban development. The nineteenth century coffee-producing and export economies of Central America provided few multipliers to stimulate significant primate city growth. Nor did the banana export economy add noticeably to urban development since most of the linkages were with other parts of the Standard or United Fruit corporate structures in the United States. It has only been in the past 30 years, with industrialization in the urban areas and the expulsion of small peasant producers, that migration to the principal cities has brought about a rapid primate city growth (see Chapter 4).

Under strategy 2, whereby the state endeavours in collaboration with the national bourgeoisie to reduce the expatriation of surplus through policies promoting local ownership and the consumption of nationally produced consumer goods, rather similar urban patterns result in spite of efforts to encourage secondary centres. Brazil, Malaysia, Tanzania, India, Mexico, Argentina, Chile and Uruguay (see Chapter 4) are contemporary illustrations of this trend. The case of Argentina is particularly interesting, given its long history of relatively autonomous development. For this reason we devote a more extensive discussion in the following section to its evolution since the late nineteenth century.

The examples of strategy 3 are either relatively rare in the Third World or so recent that it is difficult to generalize. Examples of regional decentralization can be found in Tanzania, Vietnam and Cambodia (under extreme conditions), and Cuba (with the green belt of Havana); it is also possible that rural/regional development policies in countries such as Nicaragua, Mozambique and Angola may result in changing urban-rural relationships in the future.

Internal and external factors in development: a case study

Some of the themes and issues of previous sections have been the subject of an historical study by one of the authors. Investigation of the social evolution of Argentina (in a comparison with two 'dominion' capitalist societies, Canada and Australia) in the period 1870 to 1930 indicates the complexity of the development process in a period of change (Armstrong 1980, 1981a; Armstrong and Bradbury 1983; Ehrensaft and Armstrong 1978, 1981). But one issue is clear: we cannot attribute the dramatic transformation of these societies merely, or even primarily, to external factors – an expanding international economy, European penetration, or free trade policies – although these all played a part.

Neither Argentina nor Canada and Australia could be described at this time as members of either centre or periphery in the global system. Together, they formed a neo-colonial extension of European imperial expansion; as producers of foodstuffs and raw materials for industrial capitalist markets it might be easy to categorize them as 'hewers of wood and drawers of water', typical of peripheral economies. Yet they achieved living standards, measured by literacy, income levels, birth and death rates and urbanization, comparable with those of western Europe. Indeed, these societies surpassed the achievement of that 'centre' in many respects (Clark 1951). And their strategies of domestic industrialization, however partial the results, indicated, to borrow Amin's phrase, a capacity for autocentric capitalist development.

Traditional explanations for this transition from strict specialization around a staples export sector to diversified indigenous development have focused on the favourable international economic situation in which such societies found themselves because of the rapidly growing demand for their exports. Nurkse (1962) described the international economy of the time, with its expanding trade and investment flows, as an 'engine of growth' for these fortunate participants and other international trade theorists have agreed with this interpretation.

Yet other countries' exports, tropical products and minerals in particular, were also in high demand in the late nineteenth and early twentieth centuries without generating the same patterns of domestic diversification and growth. Whereas rapid export-oriented growth added to the stock of capital available for domestic investment in the 'dominion' societies, most of the surplus from export expansion in the enclave economies was expatriated for capital accumulation elsewhere. The export sector, the basis for domestic linkages in the one case, became the source of overseas leakage in the other.

The answers to the question 'why and how did capitalist development, and especially modern domestic industrialization, take place in these three societies?' seem to lie primarily in the study of internal social structures and class relationships. Above all, the three societies were lands of European settlement. The decimation, subjugation and/or assimilation of the indigenous peoples, together with the expropriation of their lands throughout the late nineteenth century cleared away most of the precapitalist obstacles to the modernization of each economy. The stage was thus left free for the colonists who brought with them European technology, capital, institutions, social organization and the attitudes and behaviour patterns of industrial society to combine with the great natural wealth of their new lands in strategies of modernized growth.

This was the foundation upon which the class structures of each society emerged. The most powerful group was a ruling coalition of producer and trading and financial interests associated with the export sector. Because of its relative autonomy *vis-à-vis* foreign interests, this group was able to claim an important part of the nation's economic surplus for its own purposes. Some was used non-productively, but much was appropriated for investment in export sector activity and the multiplier effects spread into the rest of the economy. The coalition evolved strong and influential financial institutions to add to domestic capital accumulation with overseas borrowing and, as might be expected, shaped and used the national state to promote its class interests.

The urban working class was the other major class to appear. European migrants arriving in the three societies, and settling, for the most part, in the rapidly growing cities, also brought with them the organizations and attitudes of a European proletariat which was passing through the traumas of industrialization; the imprints of social discontent and confrontation remained with the migrants in their new countries. These urban workers gave the three societies their particular character as much as did the ruling coalition. Their equivalents were not to be found in other colonial or neo-colonial outgrowths of European imperial expansion.

The immigrants came, for the most part, as free wage earners, selling their labour on the market in exchange for commodities. In Argentina, even more than in the two other societies of European settlement, there were few opportunities to move on to land already concentrated in the hands of the large-scale *estancieros*. Consequently there was little possibility for subsistence outside the market.

Moreover, established in the urban areas, they used their literacy, labour organizations and bargaining power, at a time when the demand for labour was high, to wring occasional concessions from both employers and governments. In this way they managed to divert some of the surplus from rapidly expanding economies into improved wages and working conditions, so that by the 1920s they had achieved living standards above those of many western European societies.

A third social group comprised small businessmen, traders, store owners, manufacturers and those responsible for the smooth functioning of the modern economy and state: the professionals, managers and public administrators who supplied the technocratic and bureaucratic expertise to implement the strategies of the ruling coalition and national state.

Such structures, which, in many ways, prefigured the more recent models suggested by Poulantzas and Petras, created the conditions for a relatively coherent economy and indigenous diversified development. One consequence was the growth of a local manufacturing sector and the partial industrialization of the economy. But there are two caveats to this. In the first place, the autonomy was relative. Above all, indigenous diversification and development were always conditioned by the requirements of the dominant export sector and the largely accepted specialized role of each country in the international division of labour. Second, important divergences arose in the development patterns of the three countries as a result of their different class structures and the nature of the relationships among the major social groups. Such divergences widened as the twentieth century progressed; in the case of Argentina, especially, it seems that the degree of relative autonomy and freedom of manoeuvre has decreased with the passage of time.

The significance of this somewhat extended illustration is twofold. First, the primary basis for understanding the nature of a society's evolution and its potential for change must centre on its current internal situation set in the context of its historical experience. Centre/periphery generalizations are of value in fixing the international setting and establishing first approximations about the uneven global nature of capitalism. Consequently we believe in research of internal social structures and class relationships, since these are the prime movers in the development process.

This in no way implies that external forces – classical imperialism, neo-colonialism or the operations of transnational corporate capital – are insignificant; the evidence is clearly to the contrary. What we have sought is an approach which allows us to study the interaction of the internal and the external so as to understand the political, social, economic and cultural 'chemistry' of the relationships which condition development in each case. But, in this, the understanding of social structures and relationships *within* the society is of paramount importance.

Conclusion

Over the past decade or so both authors have been trying to clarify and refine their understanding of the processes of development and underdevelopment. Practical experience in peripheral capitalist societies has been strengthened by periods of residence and work in countries as diverse as Hong Kong, Chile, Ecuador and New Hebrides. This has broadened our understanding not only of the nature of capitalism and its functioning at a global level, but has deepened our awareness of the causes of unequal development in a number of very different societies. It is this experience which has allowed us to reformulate our ideas on the relationship between internal and external elements in the development process.

The subjection of theoretical constructs to the test of experience has, we think, given greater flexibility in approaching the complexities of national and international development. It has encouraged us to the realization that, with the flood of conceptual approaches and theoretical models offered in development studies over the past 15 years, there can be few occasions on which an absolute position can be taken. The ongoing debates have helped to widen our knowledge of social and historical processes; nevertheless much of the interchange among the dependency, underdevelopment and structuralist schools has, in a Manichaean way, confused the discussion of the problems of inequality, poverty and exploitation.

What now appears necessary is reflection upon the debates and a reappraisal of the contributions made. This will allow us to elaborate a synthesis, drawing upon the constructive aspects of the different models, and testing them against practical experience, so strengthening our theoretical and empirical understanding. The earlier quotation from Radice (page 23) provides a useful basis for re-examination: capital accumulation, the root of social inequality under capitalism, is not merely a matter of commodity and financial flows, or of distinct systems of production; it is all these, and is also concerned with political institutions, cultural behaviour patterns and values, class relations and structures and individual motivations and activity.

It is in this context that certain dependency and underdevelopment theory, in concentrating principally on exchange relations among nations as its explanation of underdevelopment, treats only part of the issue. Many critics have taken the *dependentistas* to task for their omissions and neo-Marxist approach. But not all of the dependency school has limited itself in this way; later writers have strengthened the argument by their enquiries into internal class structures and relationships. Even without this, however, a debt is owed to those *dependentistas* who, in the 1950s and 1960s, brought back a holistic and historical perspective, encouraging a more complete examination of the capitalist system than has been possible under much of conventional development analysis.

Building in part on that legacy and the critiques it generated, the modes of production and structuralist schools undertook the examination of the internal framework of capitalist society, in the modes of production, class formations and relationships, the role of the state and the articulation of all these elements. The change in emphasis was a valuable complement to dependency's concern with external relationships, for, as Castells has written,

> The excesses of the Althusserian language do not undermine the relevance of the concepts advanced and of the theoretical laws based on the Marxist tradition and on the practical experience that has nourished this theory. (Castells 1977, ix)

Nevertheless, the mechanistic quality and the economism which constrict much of the analysis, and the increasingly rarified abstraction of the model-building of the later structuralists created openings for further reformulation. Summing up the criticisms of structuralism by citing Banaji (1977), Henfrey observes that:

> despite its economism, the modes of production analysis of even economic relations is limited by its formalism – its concern with structure rather than

> movement, or 'modes' as distinct from accumulation within the capitalist mode, which will govern the effects of articulation and also vary from one case to another. For both reasons (its economism and its formalization of the economic) the modes of production approach to class struggles tends to be limited a priori to its generalized if valid statements on the role of the state and the popular sectors. (Henfrey 1981, 47–8)

With their intervention in the late 1970s, the socialist-humanist historians, too, criticized the structuralist tendency to explain social relationships and change as arising out of the inherent logic of capitalism. Instead, they highlight those elements of societal evolution through history which reflect the fluidity and subtleties of individuals and social classes, rather than structural or systemic overdetermination. This approach has restored a sense of primacy to human agency and the conscious participation of individuals and groups in the shaping of their societies, in contrast to the more impersonal, even mechanical, analyses of structuralism (Thompson 1978; Connell and Irving 1980; and the debate among Hall, Thompson and Johnson 1981). In so doing, the advocates of socialist-humanism may cite Marx in the *Third Thesis on Feuerbach*, where he wrote:

> The materialist doctrine that men are products of circumstances and upbringing and that, therefore, changed men are products of other circumstances and changed upbringing, forgets that circumstances are changed precisely by men and that the educator must himself be educated. (in Tucker 1978, 144)

We have followed these debates and our own work owes a considerable debt to the stimulus of much of the interchange. It seems to us that the attempt to find a synthesis from the florescence of ideas and approaches of the past two decades is of such importance that we follow Munck in his call for a productive research programme which avoids the competitive confrontations of 'dragon slayers' out to smash their opponents and liquidate erroneous views (Munck 1981, 176).

On the issue of internal-external determinants or conditions of development in particular we have found some recent commentaries especially valuable; Johnson's attempt to avoid the mechanical confrontation is notable in its clarity:

> If I could avoid the language of 'internal' and 'external' I surely would. It is a question of the interrelation of general determinants (the external) with specific determinants (the internal) of particular phenomena (such as militarized forms of the state) in concrete, historical circumstances.
>
> (D. L. Johnson 1981, 115–16)

He suggests that the general process of world capitalist development finds concrete expression in specific class relations or forms of the state; or better, to emphasize the dialectical relationship, he cites Palma (1978):

> The system of 'external domination' reappears as an 'internal' phenomenon through the social practices of local groups and classes, who share its interests and values. Other internal groups and forces oppose this domination, and in the concrete development of these contradictions the specific dynamic of the society is generated. (D. L. Johnson 1981, 116)

Accepting this, we have nevertheless taken a position in our studies in which the prime motivation in social change arises from internal relationships of classes and the actions of social groups and individuals in given historical situations. We begin, therefore, with the study of these structures and relationships and their implications for capital accumulation and modernization, as the key to an understanding of the causes of inequality in national development. In the following chapter we take up some of these issues which have emerged in general development theory. Our objective will be to concentrate on Third World cities and their role as the principal arenas in which these processes are played out. It is a task which has not been sufficiently considered, as we argued earlier in the chapter, but is one which, we believe, allows us to explore the many faces of dependence, structural inequality and poverty in a way which takes account of their diversity and complexity.

3

Cities: theatres of accumulation, centres of diffusion

Introduction

With the themes of the previous chapter as our general framework we are now in a position to discuss the place of cities in relation to capitalist accumulation. Cities are, stated simply, the crucial elements in accumulation at all levels, regional, national and international, providing both the institutional framework and the *locus operandi* for transnationals, local oligopoly capital and the modernizing national state.

In offering advantageous conditions for capital, cities, particularly the large metropolitan areas, act as the central places for a process leading to an increasing concentration of financial, commercial and industrial power and decision making. On the other hand, cities also play the role of diffusers of the lifestyles, customs, tastes, fashions and consumer habits of modern industrial society. These two processes of centralization and diffusion are, of course, not contradictory for the expansion of markets is a necessary part of growth in the capitalist system.

To elaborate: we see Third World urban systems not only as theatres of accumulation, but also as centres from which are diffused the culture and values of westernization. They act simultaneously as centres of operations for modern commerce, finance and industrial activity, and the providers of appropriate environments for capital's expansion and deepening. Cities are the arenas in which foreign and local capital market, advertise and sell the philosophy of modernization, efficiency and growth through imitative lifestyles and consumerism and, in so doing, undermine non-capitalist production systems and cultural values. In this sense, diffusion is a further means to enhance and promote the end of capital accumulation.

It is important to be aware of the global, historical context in which the urban systems of the Third World have evolved. The character and functions of each city have corresponded to its specific relationship with different phases of global expansion, and in particular with the capitalist system as it has unfolded over the past two centuries. Every urban area, therefore, has its own pace and trajectory of development which flow from the interaction between its internal evolution and external forces.

The trend toward industrialization is the latest phase in their historical development in many Third World cities. Earlier phases cast these cities in the role of central places for colonial or neo-colonial commercial and financial capital accumulation; many still retain this character, but others have moved into the era of capitalist industrial modernization.

In the current situation of dependent industrialization many local ruling coalitions, based in the metropolitan areas, have acted as pump primers and collaborators in the modernization of the local economy. Some of this class have invested in modern factory activity, but more often they are intermediaries, speculating in urban property, trade and financial activity, while leaving the main responsibility for industrialization to foreign capital. The diversification of ruling class interests to meet changing conditions of capitalism's expansion represents new areas of activity for Third World economies. But whether it implies a social transformation – the emergence of a national industrial bourgeoisie capable of challenging traditional oligarchies and transnational firms – is in most cases a different question. As one writer has argued in the case of Ecuador's modernization experiments of the 1970s, 'national' industrialization has only deepened the degree of penetration, especially of the urban economy, by foreign capital (Bocco 1982, 55).

With the emergence and consolidation of this latest phase of modernization, the nation state has been drawn in to a more active role in economic policy making. The promotion of industrial growth strategies has led to the channelling of resources into urban manufacturing, construction and commercial growth, and the transfer of social, regional and sectoral surplus into the modern, urban corporate sector. The principal beneficiaries of policies of subsidies, cheap credit and tax exemptions have been large national firms and TNCs. And, despite appearances to the contrary, the monetarist 'hands off' strategies employed in countries such as Chile and Argentina also represent a form of state intervention; free market policies have brought about resource reallocations in favour of national oligopolies and foreign investors at the expense of virtually every other sector and social group.

Many governments have committed themselves formally to redistributive policies with the proclaimed intention of helping low income groups, stimulating lagging regions and reactivating backward sectors. The results have generally been minimal; some urban workers in the modern sector have benefited, there has been spotty growth pole development, and parts of the commercial agricultural sector have been stimulated – though rarely those of small-scale basic food crop production – all at the expense of growing exploitation of the landless rural workers.

The dynamic of capitalist modernization has led in particular to concentrated urban-industrial expansion. In many countries the pace of metropolitan population and economic growth has accelerated compared with the national average. In Ecuador, for example, the local owners of capital climbed on the bandwagon of modernization so that

> income distribution was skewed to benefit the urban sectors, and the combination of migration, urbanization and industrialization further deepened the economic and social differences that have been the outstanding characteristic of the Ecuadorian past. (IBRD Report, cited in Bocco 1982, 45)

Even where policies of decentralization have been implemented, we find that they have rarely been relevant to the socio-economic circumstances of the area and, in particular, to its employment needs. As Gilbert observes:

> Inappropriate regional policies can increase intraregional income disparities and, as the experience of northeast Brazil demonstrates, they can be associated with absolute declines in per-capita income among the poor. (Gilbert 1980, 108)

In many Third World societies, therefore, regional and social wealth differentials have widened (see Griffin and Khan 1977) with the concentration of corporate investment in what is clearly the most profitable area for capital investment: urban-based consumer production, agribusiness and its urban processing ancillaries, resource exploitation, urban construction and real estate speculation.[1]

Skewed development and urbanization

It is this warped model of development which is responsible for what Weeks (1971) has described as the 'urban imbalance' of many dependent capitalist societies. It is a phenomenon accelerated by the concentrated patterns of transnational and national corporate capital accumulation in combination with the modernization strategies of Third World governments – whatever form these might take.

Thus the frequently enumerated problems of this urban imbalance (un- and under-employment despite the often rapid growth of the modern corporate sector, inadequate provision of housing for the poor, and the existence of a large and growing number of urban marginalized groups) are a reflection of developmental strategies based on the imperatives of capitalist sector expansion, rather than of the population explosion or of massive rural-urban migration. In fact, it is clear that migration and population growth are also outcomes of corporate expansion associated with the interests of the domestic ruling groups.[2]

Much of the argument of the preceding paragraphs is without doubt not acceptable to conventional planning wisdom which assumes that urban systems are, or can be converted into, diffusers of modern growth and development. While this may happen on occasions, the far more important role of Third World urban systems is that of bases for centralized capital accumulation by transnational firms and local businesses.

The major share of surplus produced outside the metropolitan areas is channelled to the primate cities, but a percentage will also be siphoned off to regional centres as a minority share for local investors. Part of this will go to personal consumption, luxury commodities, travel and expatriated savings; some finds its way into real estate speculation and in growing commercial, financial and, occasionally, even industrial activity; the remainder, appropriated in taxes by the state, may be used to improve the local infrastructure to attract investment and for the more profitable functioning of capital.

But a large portion is appropriated by national and international capitalists with their headquarters in the national metropoles. The modern capital-intensive sector benefits from levels of surplus-value extraction combined with the external economies provided to urban areas, and by incentives offered by the state. This becomes a prime factor in increasing the rate of profit for corporate capital in Third World countries.

Banks and other financial institutions are central to this activity[3] for they supply the servicing facilities necessary for the profitable operation of the system. It is through the financial network that the various forms of surplus, including local savings, are funnelled towards the modern urban sector (see Barratt-Brown 1974). The banking system is also one intermediary through which loans are raised on international financial markets for the modernization programmes, and is thus at the centre of the web of growing international indebtedness of many Third World nations.

The most dynamic element in this process of accumulation, on which we have touched only implicitly, is the increasing importance of transnational corporate investment (often in coalition with national capital) in the urban manufacturing sectors of Third World economies. For instance, in the three most industrialized Latin America economies (Brazil, Mexico and Argentina), data presented by Cardoso shows that United States direct investment in manufacturing more than doubled between 1946 and 1968, and the trend has continued through the 1970s into the 1980s (as will be discussed in Chapter 4), so increasing the degree of transnational control over the most dynamic sectors of the continent's economies (Cardoso 1977, in Slater 1979, 101). European investment in Third World industrialization has also grown significantly while Japanese investment in industry in the Third World, particularly Asia, has increased dramatically, as is shown in Chapter 5.

There is a transformation of such signficance in certain Third World countries that writers have been moved to describe it as the emergence of a new international division of labour.

In part, this change has arisen from the availability of cheap labour, and a more amenable work force (often composed of young women without experience of industrial or, indeed, urban society) which means that corporations can reduce production costs in their competition for overseas markets. The change has also occurred because of the opening-up of internal markets in Third World countries, where consumer demand for the products of modern industry – clothing, footwear, domestic electrical goods, building materials and plastics – has been growing steadily among different income levels. It is important to examine both these elements in the process of industrial change.

Fröbel, Heinrichs and Kreye argue that in the 1960s,

> capital almost regardless of size, has started, for the sake of survival, to reorganize transnational reproduction according to the changed conditions of capitalist expansion and accumulation. The organization of production according to the world-wide supply of disposable cheap labour, however unequally distributed, figures foremost.
>
> (1977, 84; see also 1980; also Palloix 1977)

Slater (1978, 1979) suggests at least three factors which have aided this international relocation of industrial production. First, modernization of the means of transport and reductions in the cost of shipping commodities have reduced circulation costs and made it possible to relocate some types of industry away from the global centres. Second, technological developments in the production process have allowed the simplification of some processes involving wage labour which is more willing to work long hours on monotonous tasks. Third, the costs of wage labour are generally much cheaper in the Third World, and labour is much less well organized in unions and collective bargaining units.[4]

This last factor favours alliances with local ruling classes and, when other means fail, with coercive military regimes, which can impose restrictions on the wages and working conditions of the working class in ways that are usually not acceptable in the advanced industrial societies. Fröbel *et al.* give special emphasis to the social and political conditions that capital meets or is able to impose or to 'encourage' in Third World societies. The role of the state is crucial in dealing with labour to ensure an acceptable level of free movement for capital, and provide subsidies, infrastructure and labour force training to prepare a satisfactory environment for modern manufacturing.

> There is ample evidence that the relocation of production occurs, above all, into those countries which offer total freedom in the movement of capital and which suppress all labour organization and strikes. In many parts of the world these conditions are increasingly enforced by growing militarization.
>
> (Fröbel *et al.* 1977, 85)

Another significant aspect of this pattern of industrial growth is the emergence of Free Trade Zones; more than eighty are now operating in the Third World, with almost half in Asia.[5] The growth of Free Trade or Export Production Zones is a phenomenon which traces its postwar origins back to the establishment of the designated area around Shannon's international airport in the Irish Republic in the 1950s. Since then, it has become a favourite, though not exclusive, instrument in the export-oriented strategies of the newly industrializing countries of the Third World. There is a detailed discussion of the Free Trade Zone in Chapters 4 and 9.

The development of these modern industrial sectors has also been associated with the increased production of consumer durables for the domestic markets of the Third World countries themselves. This is particularly the case in such countries as Brazil and India, but is occurring to a lesser degree in Mexico, Taiwan and South Korea. Roberts (1978) has argued that this aspect of the industrialization process is especially advanced in those Latin American societies where the process of capitalist industrialization has been in existence since the 1920s or earlier. This has involved the opening up of urban consumer markets especially to modern products, but in recent years rural consumers have also been drawn into their orbit. Here is one important reason why many Latin American nations differ so radically from Asian and African countries where small-scale 'traditional' production remains an important supplier of goods to low income populations – although the latter pattern can still be found in those societies of Latin America in which capitalist production is less advanced.

Clearly, there continues to exist great unevenness in the international distribution of modern industrial sector activity. The control and ownership of capitalist production remains highly concentrated both socially and spatially. On the other hand, there is a vigorous effort to expand and diffuse the consumption of modern sector goods. This has been typical of the capital-deepening process in industrial capitalist societies and among the élites of Third World societies for generations, but the market is now being spread beyond these groups. Carlos Filgueira (1981) describes the characteristics of this consumer expansion in the southern cone countries of Latin America as follows:

a) there is a spectacular growth of modern goods consumption, especially of durable goods, material objects and 'sophisticated' products, the major part of them being imported.

b) this is associated with a diffusion of modern lifestyles from the élites to middle and lower income groups.

c) there follows an unbalance in the allocation of expenditures on basic goods, especially food and clothing, and a deterioration in the family 'food basket' of the poor.

d) these changes are taking place within a regressive income distribution process and a growing concentration of wealth, itself the outcome of growth strategies to promote economic modernization and stabilization.

The new consumption patterns reflect changes in economic behaviour, in social relationships and the promotion of new ideological goals aimed at material success for the individual. 'Imitation is the main stimulus for action and consumption is the indicator of the success achieved' (Filgueira 1981, 108). In all this, the urban bourgeoisie plays a crucial part by virtue of its position as controllers and owners of the media and the distributive means which promote and make practicable the diffusion of new patterns of consumption throughout the society. In certain cases, of course, the practical appeal of the new products is evident; under the pressures of urban life, processed foods and household goods ranging from plastic containers to electrical equipment often help to save time which is needed for wage labour. Filgueira also notes that, from a political viewpoint, 'the expansion of consumerism can come to play an important role in the legitimation of the system' (1981, 110). It may, in turn, however, also promote increased social inequality and rural-urban imbalance; both these have the potential to create unsatisfied felt needs, frustration, growing tensions and conflicts within societies by stimulating expectations they cannot meet.

As we suggested in Chapter 1, it is clear that there is no uniformity of expansion among the spheres of production, circulation and consumption. Both the dependency and modes of production explanations provide only partial models of capitalist expansion because they focus excessively on one or other of these spheres: dependency upon circulation, the modes of production approach and, obviously, on production. Central to our thesis is the premiss that all three spheres form integral and interdependent parts of a system. This said, we lay emphasis on the need to examine the sphere of consumption in modernization. The enormous leap in transport and information technology has greatly helped international and national corporate sectors to expand global ideological influence, so creating new needs whose persuasive effects permeate the societies of the Third World.

It is our view that there is a growing similarity in the consumption patterns of Third World societies, a similarity most evident in the built environments, transport systems and lifestyles (particularly in food and dress) of the cities, but also increasingly in the regions and rural areas. Finally, it is important to note that these changes in consumption are more dramatic, both in scope and pace, than was the case historically in the advanced capitalist societies of today.

Convergence and divergence in Third World urbanization

Convergence and divergence in Third World urbanization have important implications for the character of the urban systems of Third World countries. There appear to be two contradictory developments. On one hand there is a growing *heterogeneity* among Third World countries in terms of their economic activities. The specific variations in the international expansion of capital make it necessary to distinguish the industrializing export economies such as Hong Kong, Taiwan, South Korea or Singapore, and larger internally directed industrialized countries such as Mexico, India and Brazil, from raw-material exporting countries such as Zaire or Nigeria and the low income agricultural exporters such as Bangladesh.

On the other hand, while each society has its own specific experience in its relationships with the international capitalist system, there seems, especially in the large cities, to be a growing *convergence* of urbanization patterns (see Roberts 1978, Browning and Roberts 1980) as a result of the impact of twentieth-century corporate investment. This is particularly applicable to Latin America where, according to Roberts, there is a continent-wide convergence of urbanization patterns (Browning and Roberts 1980, 97). In fact, these apparently contradictory patterns of heterogeneity and convergence are a reflection of the uneven penetration of investment and the diverse forms of capital found in Third World societies.

In many underdeveloped countries, patterns of primate city growth have followed those of the global cities of the industrial capitalist nations in certain respects – in providing central places for transnational corporations, world financial institutions and business, administrative and political élites. As such, they too have evolved modern economic sectors and class structures, and a built environment comparable with those of the global metropolises such as New York, Tokyo and Frankfurt (see Friedmann and Wolf 1982).

But divergent forces are also at work within Third World societies. The differential operations of international and national capital operation and the particular historical conjuncture mean that the convergence theory cannot easily be applied to other parts of the national urban system where capitalist modernization is less pronounced. Kamal Salih argues that regional disparities, therefore, arise

> largely because the region in the periphery . . . is twice removed from the impact of the transnationalization process. This is in the immediate sense that it is first 'mediated' by its relationship to the national core region in the spatial division of labour, and then by the relation of the national economy in the international division of labour. (Salih 1981, 15–16)

However, with the modernization of even the remotest areas, certain aspects of the lifestyles and behaviour patterns (including the consumerism) of the industrial societies are adopted in market towns and villages even though their production structures may be totally undeveloped. This understanding of the uneven impact of capital on both production and consumption patterns is therefore central in explaining the quite bewildering mixes of primacy and urban balance which characterize the urban hierarchy of Third World countries.

The trend towards convergence is creating a ruling class with international values based in metropolitan areas. This has been greatly facilitated by the technological improvements in distribution (transport systems, warehousing and retailing innovations) and information (the use of the media and advertising to mould public opinion) and by the diffusion of selective information. The consequence has been to transform Third World metropolitan centres into producers and providers of goods and services for the rest of the country. To ensure the profitability of such activity the mass media play their role by spreading the message of modernization to the rest of the nation. The result is a dissemination by advertising of new cultural and behavioural patterns which encourage the consumption of modern goods and services, even though the productive means to supply these inculcated needs are not usually locally available.

What conclusions can we draw from these developments? The first is that there are, simultaneously, trends towards both convergence and divergence. On the one hand this shows up in the growth of derivative modern production facilities in the metropolises of Third World countries. But this type of convergence on the models provided by the global cities such as New York, Frankfurt and Tokyo is matched by a growing divergence *within* Third World societies between their primate cities and their regions and rural areas.

Another form of convergence is also taking place in the sphere of consumption within the Third World society as the national centres, themselves recipients of imitative consumption patterns and imported lifestyles, transmit these, in turn, to smaller centres and rural areas. The primate city, therefore, plays the role both of centralizer of productive and distributive functions and diffuser of modernization ideology, opening new markets and stimulating consumer demand. The structure emerging in Third World societies is one of dispersion of demand (consumption) and the centralization of supply (production); it is a situation in which the diffusion activities of the metropolitan groups, in fact, result in the further concentration of their economic, social, cultural and political power.

One should not neglect the role of some dynamic regional centres in this process. Thus in Mexico, Monterrey and Guadalajara, and in Malaysia, Penang and Johore Bahru act as alternative centres of growth and dispersion to the primate cities.

Following on from Filgueira, we can see that the promotion of modern consumerism presents another face of capital accumulation; it contributes capital-deepening by the corrosion of traditional values and their replacement by the prescriptions of the market. The diffusion of modern lifestyles, through mass media and advertising, wears down the older social, cultural and economic systems and creates clienteles ready for the products of the urban capitalist sector.

We have attempted to illustrate the main features of these processes in Figure 3.1. This combines the arguments of this section with the hierarchical model of TNC organization suggested by Hymer (1975), who divides corporate decision-making functions into Level I (global), Level II (peripheral co-ordinating cities) and Level III (day-to-day production in the field). Speaking of Level II activities, Hymer notes that:

> because of their need for white-collar workers, communications systems and information [they] tend to concentrate in large cities. Since their demands are similar, corporations from different industries tend to place their coordinating offices in the same city, and Level II activities are consequently far more geographically concentrated than Level III activities. (1975, 50)

The diagram thus illustrates how decision making, surplus transfers and capital accumulation are centralized at different levels while ideology and behaviour patterns, together with policy decisions in particular, are diffused to the regional and rural areas. Once again, rather than describing just spatial relationship, we are concerned with a series of socio-economic interactions among different groups – corporate businessmen, bankers, local merchants, craftspeople, peasants and wage-workers – found in various parts of the Third World urban/regional/rural hierarchy.

The policy implications of seeing cities as centres for the accumulation of locally derived surplus for investment by foreign and local enterprise make it necessary to analyse in detail the distinctive features of capitalism. This is particularly germane to any study of the role of 'urban centres' as development poles in the national economic growth process.

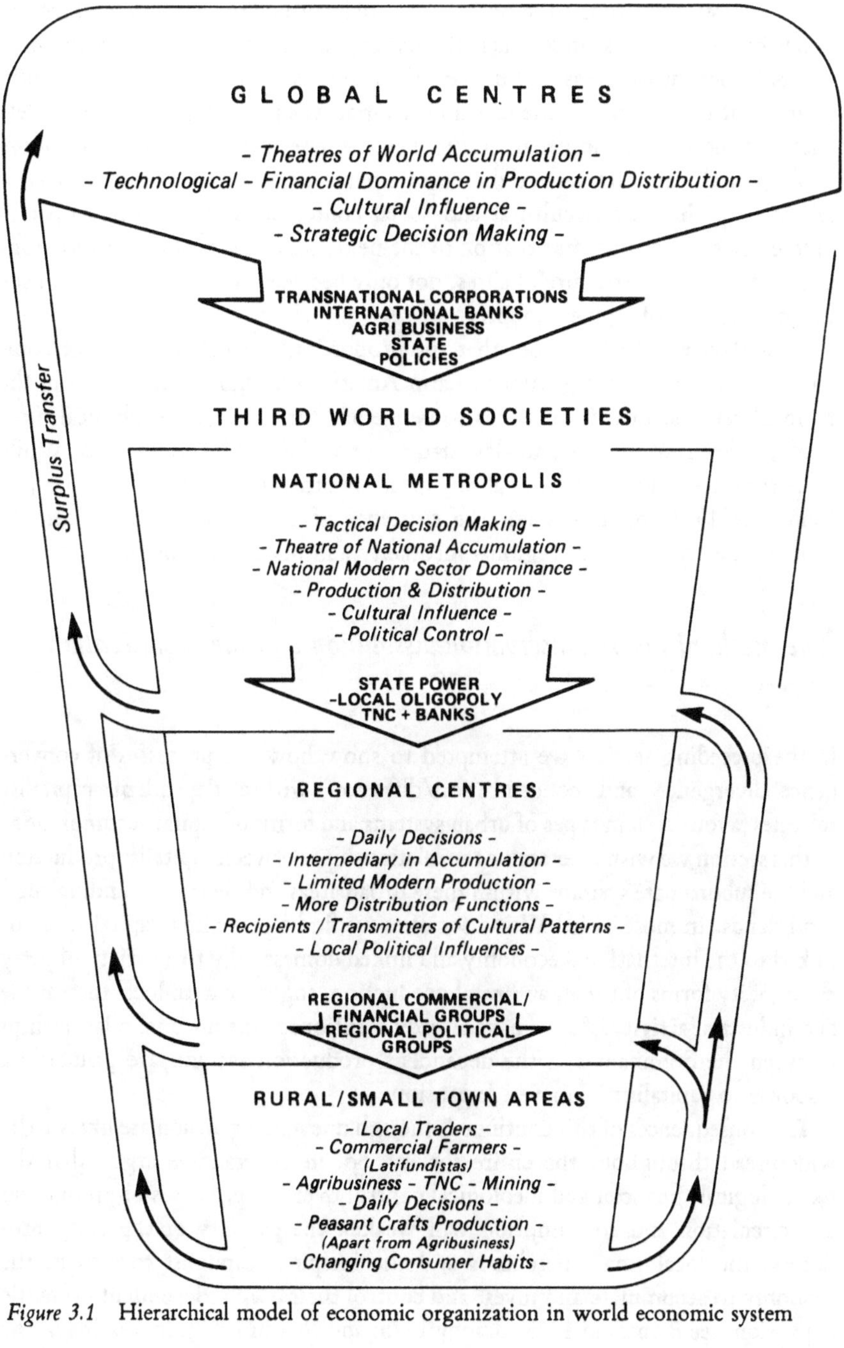

Figure 3.1 Hierarchical model of economic organization in world economic system

Up to now, primate and regional cities have furnished the base from which the urban bourgeoisie has undertaken the expropriation of surplus from the producers in non-urban areas and in the urban centres as well. This has evidently occurred at the expense of the urban proletariat, the peasantry, the rural proletariat and the village and urban craftworkers. Under capitalism the city's morphology and structure reflect the requirements of accumulation and of class relationships in various cultural and social contexts. As these forces operate differently from one urban situation to the next, we can expect to see variations in urban form and structure of cities, not only between countries but also in the urban hierarchy of a given country.

Thus Roberts (1978) argues that the proportion of small-scale simple commodity activity is much greater in Latin American secondary cities than in the national centres. In turn, this has consequences for the types of physical morphology, levels of income and class structures of the two levels of cities. While there are some doubts as to the validity of this in other Third World contexts (see Johnstone 1981, on Malaysia), the existence of the pattern would seem to support the arguments of regional inequality and underdevelopment.

The mechanisms of conservation-dissolution and the urbanization process

In the preceding section we attempted to show how the processes of convergence/divergence and concentration/diffusion within dependent capitalist societies favour certain types of urban systems and forms of capital accumulation. In this section we wish to examine the relationships between capitalist production and the subordinate systems within these urban areas and their surrounding rural hinterlands. In most Third World societies we find a prevailing capitalist sector linked to the international economy and linked domestically to a variety of petty commodity forms of urban and rural production, together with local mercantile and industrial activity. As we have already mentioned, the unequal relationships between the dominant and the dependent production systems are primarily a response to capitalism's uneven expansion.

The consequences of this continued, though uneven, expansion are likely to be widespread throughout the entire society. So, in general, we argue that the technological, financial and ideological control over the processes of production and circulation and consumption will weaken the capacity of the petty producers, the local small merchants and the incipient capitalist groups in the economy to accumulate and invest and control their own independent domestic expansion (see Bienefield 1975; Bradby 1975; and Griffin 1969). Increasingly, the development of the society comes to revolve around the decisions of the corporate state as to where, when and how – or if – it will expand economic growth so as to promote structural change.

Bettelheim argues with some force

> *that it is the world domination by the capitalist mode of production*, resulting from economic, political and ideological relations the effects of which are called colonization and imperialism, that has favoured the maintenance (or in some cases the development) in the dominated countries of production relations and also of political and ideological relations which have 'blocked' the development of the productive forces. (1972, 291)

He elaborates by referring to the process of 'conservation-dissolution', which means

> that within a capitalist social formation the non-capitalist forms of production, before they disappear, are 'restructured' (partly dissolved) and thus *subordinated* to the predominant capitalist relationships (and so *conserved*). (1972, 297)
>
> This is the aspect that is interpreted, in ideological terms, by the metaphor of the 'dualistic character' of these social formations – a metaphor that *conceals* the specific type of domination by the capitalist mode of production that is characteristic of these formations. (1972, 298)

Le Brun and Gerry have taken the argument another step in observing that:

> generally speaking it is the dissolution aspect which dominates the conservation aspect in developed countries of the capitalist mode of production. Petty production is therefore residual and tends to disappear. In Europe this dissolution mechanism operated during the period of transition, which, due to the Industrial Revolution, led to the establishment of a self-sufficient capitalist mode of production. In the countries of the periphery it seems it is the conservation aspect which dominates. (Le Brun and Gerry 1973, 9)[6]

In the light of our examination of the expansion of capital, this assertion of Le Brun and Gerry now needs qualifying. The impact of agribusiness, the green revolution strategies of TNCs, the growing investment in industry, the involvement of local capital in the acquisition of land, and the encouragement of mass consumption have now altered the situation in many Third World societies. Nevertheless the argument retains validity enough for us to ask why capitalism does not act constantly to dissolve other systems. Three reasons emerge. First, many of the activities in which the small-scale sector is engaged are still often unprofitable for capitalist production because, as Le Brun and Gerry point out, it is essentially orientated

> towards the production of exports for metropolitan consumption and/or transformation for re-export, and the production of consumption goods for the more favoured strata of the population. (1973, 10)

As will be obvious from our preceding discussion, it is remarkable how quickly this assertion has begun to crumble in the face of the rapid modernization of production, distribution and consumption in Third World countries.

The second reason, as stated by Meillasoux (1972), the persistence of a large non-capitalist sector, typified by underemployment and low income, has encouraged the growth of a cheap surplus labour force which, in turn, has helped to keep wage levels down. This sector is also available to the capitalist for putting-out operations when he wishes to cut production costs. The putting-out system may also represent a transition from non-capitalist to capitalist production. The situation is thus one of continual pressure and tension, but with the trend to dissolution increasingly in the ascendant.

Finally, the existence of a large petty production and distribution sector characterized by family employment is important to the strategies of the bureaucracies of Third World states, since it reduces the need for social welfare investment. More capital can, therefore, be made available for allocation to these activities which support and promote the expansion of the modern sector. There is also the rather more debatable issue, that the preservation of a large part of the population in labour-absorptive sectors reduces opportunities for political unrest in the large urban centres (see Nelson 1976, 1979).

To summarize, it is clearly in the interests of the local ruling classes and international capital to allow these non-capitalist systems to continue, at least in the short term, so as to avoid some of the more extreme structural contradictions which are emerging from the unequal alliance between transnational and local capital.

It should not be imagined, however, that conservation is a linear movement or takes place in any static sense. Small-scale production and distribution activities responsive to the exigencies and penetration of the prevailing system, develop, regress, disappear and are transformed. We are witness to patterns of growth and change which are involutionary, and above all, occur within the framework of constraints set by modern capitalist production, distribution and consumption. We are not dealing in any sense with a dualistic society in which the two sectors are separate, or in which the so-called 'traditional' part functions independently of the 'modern' (see Bettelheim 1972, 298). Non-capitalist forms are tolerated, or conserved for the time being, in part because it is generally in the interests of the prevailing system to allow them to continue.

An important additional reason for the continuance of the non-capitalist sector, which we only touch upon here, is the constant struggle by simple commodity producers (for example, farmers and craft producers) to retain control over their means of production. This social effort is shown by their willingness to exploit themselves and their families, to seek local support work as a family and to migrate in search of jobs; it is an impressive illustration of the resilience and resistance of those who wish to preserve their independent status. Examples of this struggle are given in the chapters on Ecuador. Elsewhere one of the authors has described such *preservation* particularly when applied to cultural persistence in certain craft industries (see McGee 1979a).

The *dissolution* process occurs at the same time for capitalist enterprises under the pressure of competition and the constant need to expand. In this way capitalism seeks to increase its rate of profit, in part from widening the market, and also from appropriating a greater volume of surplus value from the new employment of low-paid wage labour.

Dissolution, moreover, is often facilitated by the intervention of the state in the interests of 'modernizing' the economy. The encouragement of foreign enterprise, producing consumer items such as shoes or clothing, or the banning of a multitude of small-scale transport units to allow a large-scale transport company to control the public transport system, are obvious examples of interference which favour the dissolution process. More subtle are policies which establish standards of quality or hygiene controls based on standards current in the developed countries, and which, given the capital required, can only be met by large corporations with significant reserves.[7]

From this analysis, it is clear that conservation and dissolution are both *complementary* and *contradictory* processes which can only be understood within the context of global capital accumulation. Differential pressures are exerted upon subordinate systems according to the degree of penetration and modernization of the society. More specifically, conservation will be most in evidence in those countries which are still in a relatively early phase of their integration into the world capitalist system, since the nature of their involvement (agricultural or mining export activity) may require only a limited, and generally seasonal, expansion of the labour market. It is, indeed, possible that indigenous communities and their social, economic and cultural organization may not only survive but even be strengthened in this phase of contact.

Activating the theory

We have tried to represent these processes, and the role of urban systems in them, in the following Figures. Figure 3.2 is a simple representation of the world

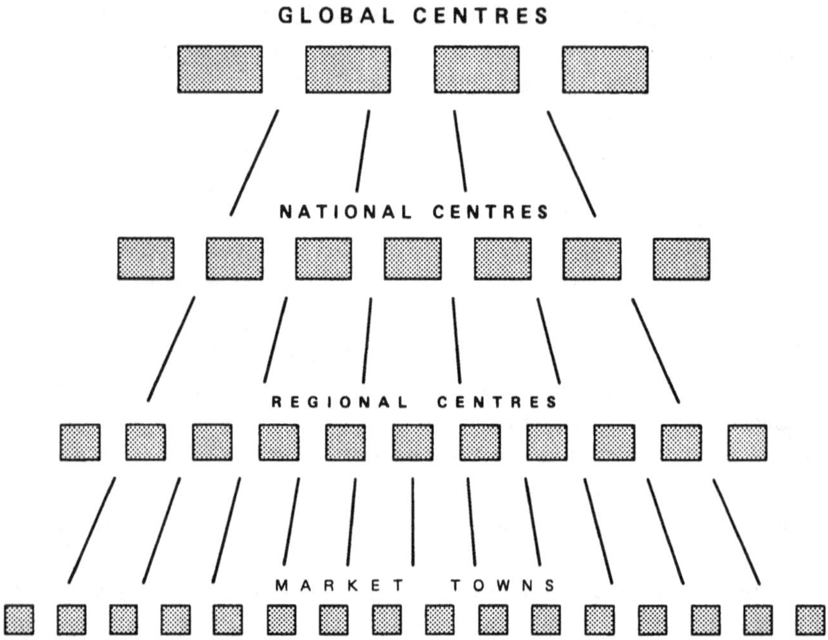

Figure 3.2 World urban hierarchy

capitalist urban hierarchy. The various components of the system are distinguished not so much by the conventional descriptive elements of urban typologies such as size, but rather in terms of their function in the accumulation process undertaken by transnational corporations and local ruling classes. Thus *global cities* such as New York, Tokyo and London, are the world's major financial centres as well as being the national headquarters for the majority of the transnational corporations. They act as global accumulative and decision-making focuses, attracting capital for further investment and expansion on an international scale. The *continental cities* exercise a similar function but within a more restricted ambit. Thus, Singapore and Hong Kong act as a major sub-accumulation centre for much of the insular Asian region, while the São-Paulo-Buenos Aires axis performs somewhat the same role within South America.[8]

The *national cities* dominate within their countries, acting both as focuses for national accumulation and as links, through transnational enterprise, to the global centres. While the national metropolises are often the capitals as well as being the largest centres in the country (such as Santiago, Chile) they may be large urban agglomerations without formal national capital functions, such as Guayaquil, Ecuador, or Surabaya in Indonesia. These national cities interact and influence regional centres and then regulate the countryside through a system of intermediary centres. Figure 3.1 (page 51) gives a more dynamic impression of the capital flows, together with those of political and ideological influence, which help to create an 'appropriate climate' for a more efficacious process of capital formation and investment by corporate enterprise.

In Figure 3.3 we have discarded the urban hierarchical framework discussed in Figure 3.2 in an effort to show more effectively the manner in which capitalism expands and engages in the process of accumulation. Such is the complexity of the world urban hierarchy that it is not possible, for example, to indicate the fit between global cities and the transnational corporate sector, although it is clear, as Cohen (1981) has shown, that transnationals use these cities as their bases. Rather than incorporate each urban level into this model of capitalist expansion depicted here, we have preferred to concentrate on capital and its operations, leaving the role of urban systems to other Figures.

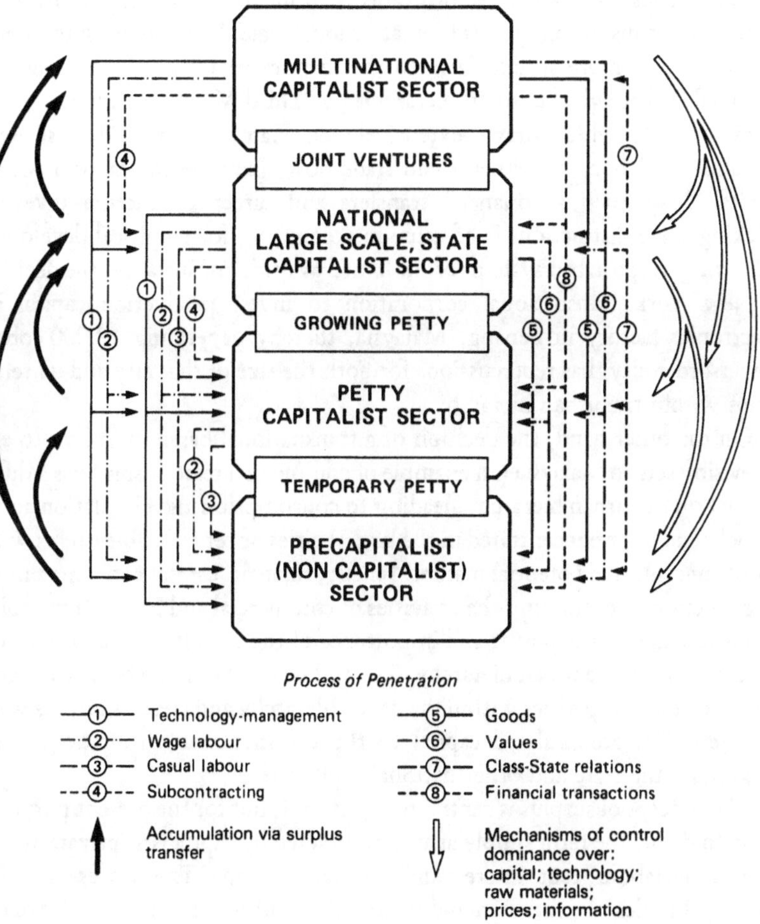

Figure 3.3 Capitalist penetration and accumulation

Figure 3.3, then, attempts to show the way in which capitalism interacts with and penetrates non-capitalist systems. Central to this is the assumption that the eight processes of penetration all operate in the spheres of production, circulation and consumption. It is the ability of the capitalist sectors, represented by transnational and national corporations, to control and dominate these relationships which is crucial to the rate at which dissolution occurs in non-capitalist activities. The diagram illustrates how penetration and accumulation involve the four sectors within the world economic system, suggesting that certain intermediary sectors such as joint ventures and the evolution of petty capitalist through to large-scale national capitalist firms emerge as a consequence. The diversity of the structures and relationships indicated in the diagram are elaborated with concrete examples in the case study chapters.

The purpose of these three diagrams is to illustrate the general processes of capitalist expansion and penetration at a global scale. We now wish to direct our attention in a more specific way to the manner in which these processes are channelled through the urban hierarchies of Third World societies.

Figure 3.4, which draws extensively on Kamal Salih (1980), shows the dynamics of labour movements and trade flows (commercial capital) and other capital flows such as financial transfers and direct production investment. Making these distinctions is of value in accounting for empirical developments occurring in the urban systems of the Third World. For instance, the decision by a New York transnational corporation to invest production capital in an electronics factory in Penang, Malaysia, thereby generating 12,000 jobs in a lower-order city, has repercussions for both the size of that city and its relationships within the urban hierarchy.

On the other hand, the decision of a transnational based in Tokyo to export television sets to Malaysia (an example of commercial relationship) has a different impact on the urban hierarchy, leading to consumption and circulation activities which tend to be concentrated in the largest cities or central sub-systems with the most concentrated potential market. Finally, financial capital performs a lubricating function in promoting the activities of commercial and production capital. It tends to enhance the power and importance of the largest cities and central sub-systems within the urban hierarchy. The end product of these flows is, of course, to increase the disparities within the urban hierarchy and create a surplus which is transferred as accumulated capital to the transnational corporation, national oligopoly, the state and other national capitalists.

The concept of surplus is central to Figure 3.4, but for the present purposes we wish to define the term simply as value extracted by capitalists (private and state) from wage labour, rent, interest and unequal exchange. This process is subject to historical cycles of expansion and contraction and is exceptionally uneven in its class and spatial impacts.

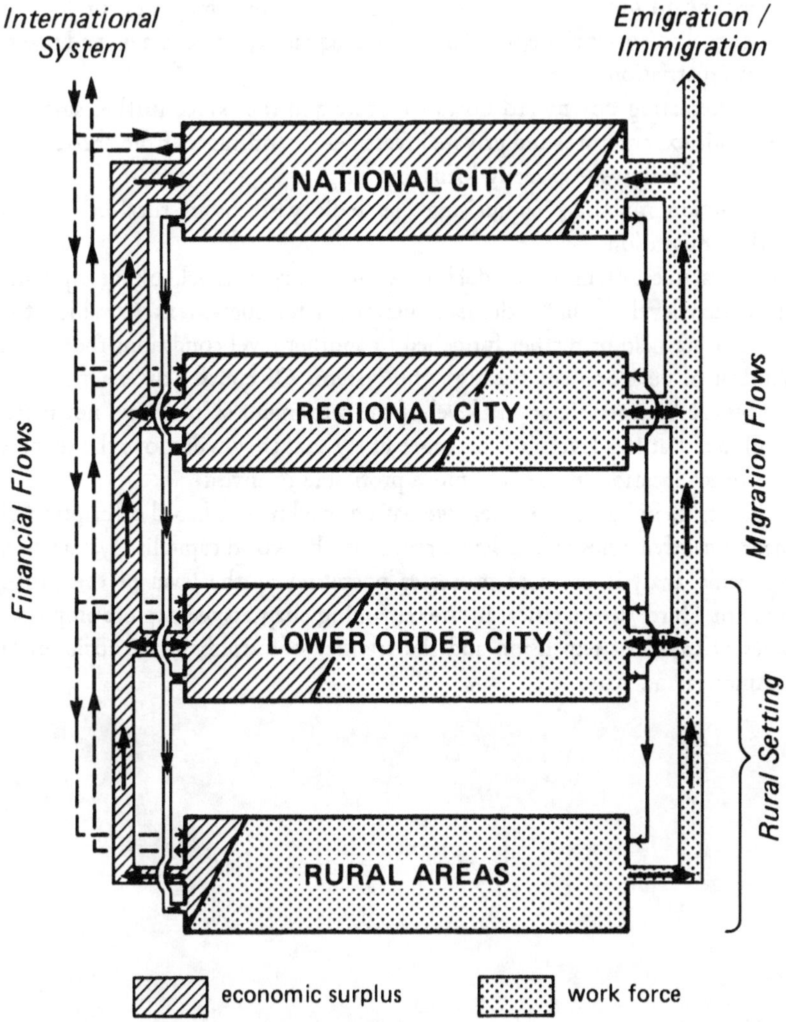

Figure 3.4 Capital labour flows and accumulation in a hypothetical urban system

Notes: Financial Flows – solid arrows indicate internal financial transactions; broken arrows indicate external financial transactions. Migration Flows – thick arrows indicate permanent migration; thin arrows indicate circular migration.

The following components have been listed as important by Salih (1980) in modelling this process of extraction:

a) The amount of the surplus appropriated by the various classes (relative distribution of the surplus is shown in the urban boxes).

b) The size and structure of the labour force divided into various classes.

c) The upward flow of surplus through the urban hierarchy and rural areas. The basic mechanisms include transfers from unequal exchange (trade) and taxes, as well as repatriation of profits.

d) Counteracting downward flows to create and reproduce further surplus, such as public expenditures and private investment, remittances from abroad and foreign capital inflows (especially loans and aid).

e) Population mobility, comprising the basic patterns of rural-urban and inter-urban migration.

The central question in this model is how much surplus is held by the capitalist sector at each level of the settlement hierarchy, for the extent to which the surplus is either held or further funnelled to another level conditions the spatial distribution of development through the system. Of course, the retention of surplus by the bourgeoisie at any level of the settlement system does not mean they will invest at this level. It is far more likely that they will move the surplus to some other location where it is more profitable to invest.

In Chapter 2, and in this chapter, we have moved from a broad description of the theoretical dynamics of the development in the world capitalist system to a more specific analysis of these processes operating at the level of the urban hierarchy of Third World cities. In the remaining chapters we explore empirical examples of capitalist expansion and the transfer of surplus in the societies of Latin America and Asia.

4 Accumulation and Latin American cities

Introduction

We begin our more empirical examination by drawing attention to two aspects of the accumulation-urbanization relationship in Latin America. The first concerns the impact since the Second World War of expanding international capitalism on the continent. The second examines the interaction between this expansion and the evolving internal social, economic and political forces and the implications for urban development.

Historically, capital accumulation in the continent has been closely associated with foreign penetration through piracy and plunder, mercantilist commerce, portfolio investment and, most recently, direct production investment (accompanied by consumption promotion) by transnational corporations. Each phase has been characterized by its own type of development, the specific patterns in each society arising out of the interactions between the internal conditions and external forces.

As an example, we can note the variations in the historical patterns of urbanization within the continent. Soon after European conquest, the imposition of imperial control was characterized by a shift of urban settlement to the coast. Colonial cities became administrative sub-centres for the empire and entrepôts for mineral and raw material exports, as well as the funnels of metropolitan products. They also acted as the central places for capital accumulation; the architectural magnificence of many colonial cities is clear evidence of the way that part of the extracted surplus was used.

Changes in patterns of urban development continued after independence in the early nineteenth century as each society responded in its own way to the expansion of the international economy. Buenos Aires and Montevideo, villages during Spanish rule, became metropolitan centres with modern ruling classes taking advantage of nineteenth-century free trade and the European demand for their temperate staples exports. Lima, a vice-regal city under the empire, gradually lost its importance with post-independence commercial changes. Similarly, following the earlier discoveries of gold in Minas Gerais, the cities of central-southern Brazil came to supplant those of the north. The southern economy then grew under the stimulus of an expanding international economy and the response to it of a dynamic ruling class using the surplus gained from coffee sales to invest in the local economy.

The association between international development and changing patterns of urban growth is evident. But the role of cities as central places for social groups capitalizing on such commercial activity is still more important. Nineteenth-century Buenos Aires became the economic and political power base for the ruling coalition of traders and bankers mediating the nation's commerce and land-owners whose *estancias* in the *pampas* produced the wheat, wool and meat for overseas markets. Later, the city provided the environment – including the linkages from the wealth-generating export sector, a centralized consumer market and a rapidly growing labour force – in which further diversification into domestic manufacturing could take place.

It is this broad-based national economic development, itself the result of a relatively autonomous evolution of social forces and political institutions (Armstrong and Bradbury 1983), which distinguishes Buenos Aires, Montevideo, Rio de Janeiro and, perhaps, Santiago from other Latin American cities of the early twentieth century. The expansion and diversification of Buenos Aires reflects the way in which Argentina profited by its position in the international economy of the late nineteenth century. Ruling class and state autonomy in the face of external pressures contrasts strikingly with other societies in the continent whose lack of local political control and economic integration resulted in external relationships marked by deeper subordination to foreign interests.

Industrial production and accumulation

Latin America has provided an attractive and growing market for modern foreign investment since the mid-nineteenth century. In the past 30 years, the changes taking place in postwar capitalism have made a profound impact on the continent: concentration and centralization of capital in the advanced industrial economies; corporate spill-over into the Third World as transnationals seek markets, resources and cheap labour; and the political, economic and cultural hegemony of the United States. Only in recent years has this ascendancy come under challenge by the corporate expansion of other industrial capitalist nations (Slater 1979, 97ff.).

How has the postwar evolution of global capitalism affected Latin America? Direct foreign investment in manufacturing has much more powerful integrative effects in the host economies than had earlier portfolio investment and commercial relationships (Palloix 1977), so that 'the process of postwar industrialization has had the special feature of generating a more rapid expansion and a closer solidarity of the system as a whole' (Tavares and Teixeira 1981, 87). But the impact has been variable. Certain countries – and regions and cities within them – have been affected more than others, and it is in these that spatial/structural unevenness assumes such marked characteristics of the modernization process, as Castells has shown in his historical comparison of the effects of European penetration of the continent in the nineteenth and twentieth centuries (Castells 1977, 59ff.).

In all this, the transnational corporations and international banks are at the leading edge of postwar capitalist expansion. The attempt to build a new world economic order under United States leadership began with the Bretton Woods Agreements towards the end of the Second World War. International economic relationships were to be stabilized, and barriers to the free flow of trade, investment and labour were to be dissolved as far as possible (Armstrong 1981a, 58–9). Members of the IMF, the World Bank and the GATT were urged to adjust their domestic policies to an open market strategy, a goal which has remained central to United States policies over the years.

Even so, within Latin America, development policies initially appeared to be taking a different path. Under the influence of the Economic Commission for Latin America economic development strategies in the 1950s and 1960s, traditional *hacia afuera* policies emphasizing growth based on raw materials and foodstuffs production for export, were to give way to more diversified internally oriented (*hacia adentro*) development. Import substitution manufacturing, promoted and protected by the state, aimed to create new internal economic and social alignments while also changing international relationships in a reconstituted world division of labour (Pinto and Kñakal 1972; ECLA 1961; Rodriguez 1979).

Over the years, however, the *hacia adentro* strategy did little either to create a progressive, nationalist manufacturing class or to change the dependent nature of the continent's links with the industrial capitalist world. The very process of import substitution, heavily reliant upon foreign investment, led to a growing dependence upon imported technology as well as the foreign corporate take-over of much new industrial activity. At the same time, the import requirements of the new industries led to external indebtedness and growing balance of payments deficits (Booth 1975, 57–61). These shortcomings opened the way later for alternative approaches. One was the radical critique of the dependency school already surveyed in Chapter 2. Of greater practical impact has been the pressure by TNCs, international banks and developmentalist regimes to return to the free market *hacia afuera* strategy.

Since the 1960s, one leg of this strategy has rested upon agricultural modernization to revolutionize *latifundio* and peasant agriculture by capital-intensive methods of production and distribution. The second involves sectoral restructuring by direct capital-intensive investment. In the past two decades United States, European and Japanese manufacturers have concentrated investment in certain favoured regions and industries. In more industrially developed economies, especially Brazil, they have consorted with the state and private capital.

Industrialized agriculture

The modernization strategy is now making the distinction between agricultural and manufacturing activity increasingly artificial. Feder (1977) has argued that the purpose of agribusiness is to commercialize agriculture and integrate it into global corporate activity. The producer – farmer or peasant – becomes but one part of the overall system which encompasses 'upstream' investment in machinery, fertilizers and pesticides, and the 'downstream' activities of processing, storage, transport and marketing (George 1976, 1979).

Even where the producer retains formal control of the land, he is sandwiched into a 'sequential capital-technology package' (Feder 1976a, 427) at the beck and call of corporate strategists deciding what will be produced, when, and how, and establishing the quality and price terms for the products. The farmer is left to enjoy the remnants of his independence, facing risks of drought, disease and the vagaries of unstable market demand.

What are the characteristics of this agrarian revolution? First, capital and technology are channelled into export-oriented production – soybeans, sorghum, specialty fruits, vegetables, flowers and meat, as well as the more traditional staples – rather than basic crops for local market consumption. Second, modernized agriculture in Latin America is not easily distinguishable from its North American counterpart in terms of capital, technology and management (Burbach and Flynn 1980, 84ff.). The one clear difference is in salary scales; relatively few workers in rural Latin America are organized to negotiate improved wages and conditions. And repression has often been the response to those who have tried.

Third, capital-technology-management transfers are pushing small owners and tenants off the land, converting many into full- or part-time rural wage workers or driving them to the cities. The modernization/green revolution/agribusiness strategy is now transforming rural social relationships into the classical confrontation of wage-worker and capitalist. (The movement is not all one way, however, and, as we have argued earlier in the conservation/dissolution discussion, many small producers are seeking ways to resist pressures and hold on to their land.) Finally, although the corporate penetration of agriculture is occurring throughout the continent, it tends to be concentrated more in the larger countries such as Brazil, Argentina, Colombia and Mexico, together with the close-at-hand Central American and Caribbean economies. Guatemala, in particular, has acted as a type of nursery for members of the Latin American Agricultural Development Group (LAAD), a consortium of United States banks, chemical, machinery and food corporations involved in agribusiness (George 1979, 46; Barry *et al.* 1982, 25ff.).

Free Trade Zones

The narrowing of the space between cities and rural areas can also be seen in the operations of the Free Trade Zones (FTZs). Zones are designated by governments in agreement with international businesses planning to diversify their factor sourcing. Local objectives have revolved around ideas of national modernization and employment creation but the results of such economic restructuring have often been to increase the dependence and vulnerability of the host country while offering the corporation cheap labour and subsidized production costs in the export enclaves (Adam 1975, 100ff.).

As with agribusiness operations, many FTZs have been established near to the North American market in the Caribbean and Central America, and, above all, along the Mexican-United States border (Fröbel, Heinrichs and Kreye 1980, 309; Girault 1981, 278–83; Flynn, Santana and Shapiro 1980, 13ff.).

What are the characteristics of these global assembly lines? In the first place, they are geared to export markets, with few links to the national economy or the needs of local consumers (in fact, the parent companies using cheap, labour-intensive output for later stages of processing elsewhere often provide their sole market). Second, they tend to be located in enclaves near ports (as in La Romana, Dominican Republic), international airports (San Bartolo, El Salvador), or in areas virtually integrated into the United States' economy (Mexico's *maquila* zone).

Third, they impose costs on the host country, as the state is required not only to invest in infrastructure, but also – as in the Mexican government's Border Industrialization Program – to provide tax and credit inducements to potential investors. Fourth, labour is sought at low wage rates (even through the TNC may often pay higher levels than local firms), and so the *maquiladora* workforce tends to be female and non-unionized. Long hours and intensive, repetitive assembly-line work with few chances of promotion are characteristic of working conditions. Fifth, government policy strives to ensure that such conditions continue. In El Salvador, for example, draconian labour policies in the 1970s restricted the right of labour to organize or to strike in the enclaves (Robert Armstrong 1980, 13; *Latin American Economic Report*, 1979a and b).

Sixth, while much of the production focuses on the manufacture of electronics, furniture and clothing, there are also examples of enclaves, as in the Dominican Republic, where 'rural' (sugar) and 'urban' forms of export production are taking place alongside each other (Girault 1981, 278–83). Distinctions between the two now seem increasingly difficult to draw under conditions of corporate modernization. Finally, as Robert Armstrong (1980) has shown in the case of El Salvador, motives for the state-promoted creation of the FTZs may vary, but an especially clear one is that of class interest. Alone, or in partnership with foreign capital, local investors enjoy the advantages of the export industrialization strategy: subsidies, tax rebates and a controlled labour force.

The FTZs with their *maquiladoras* are a result of capital's postwar global expansion, but they are only one form that corporate growth has taken in the context of the Bretton Woods system. Latin America's role in the international economic order has changed during this period. In addition to providing the raw materials, foodstuffs and migrant labour for the industrial capitalist economies, some 'new industrializing countries' (NICs) have experienced rapid growth of foreign investment in their manufacturing sectors. Such a 'reordering of the international division of labour' (Urencio 1979, 1207) stemmed initially from the decision to relocate less competitive branches of industry away from highly industrialized economies to avoid the higher costs of labour-intensive assembly operations (see also Landsberg 1979, 55ff.).

Furthermore, the Trilateral Commission strategy for industrializing certain economies in the continent, Urencio observes, has now taken root in the larger and more advanced countries. Brazil, Mexico and Argentina, especially, have been encouraged

> to carry out a strategy of externally opening their economies . . . [so that TNCs can] reorganize productive activities, especially in those branches which are in crisis throughout the world, such as automobiles, steel, textiles and certain chemical products. (Urencio 1979, 1208; also Movarec 1982)

The net effect of these strategic decisions on Latin America has been to increase industrial output over three decades by 6.7 per cent annually (compared with a world average of 5.9 per cent, 3.6 per cent in North America and 5.2 per cent in

Europe). Industrial exports have also leaped; from 1950 to 1975, steel production increased thirteen times, machinery and equipment ninefold, electrical energy eight times, cement, seven, and automobile production, almost non-existent in 1950, accounted for nearly two million units in 1975. In some of the economies the share of manufacturing in total GDP is comparable with that of the highly industrialized nations (Soza 1981, 41–2; Marinho 1981, 11).

Such growth imposes certain conditions, in particular a growing dependence upon foreign technology, management and financing (Urencio 1979; Dodero 1979; Slater 1979; Ruiz Nápoles 1981). In Brazil, for example, the attempt by the state to create a viable national public-private industrial structure (Evans 1979; Munck 1979) has been circumscribed by the processes of 'internationalization' and 'denationalization' of the manufacturing sector,[1] so that less than 30 per cent of the modern industrial sector remains under the exclusive control of national enterprise (Tavares and Teixeira 1981, 98). At the same time, the structure of a manufacturing sector which, in part, supplies export markets but is geared even more to the local market, has been an important factor in Brazil's enormous foreign debt. Heavy borrowing for direct productive and infrastructural investment, together with the outflow of hard currency in the form of profits, royalties, service charges and other payments are all elements in the country's modernization strategy[2] (Juruna 1981; *Comercio Exterior* 1981, 269–74).

The same is true for Mexico (Jenkins 1979) and Argentina (Martel 1978), so that the foreign debt for the three principal economies of the continent stands at a total of more than $220 billion. The contradictions and social tensions arising from the open market model have become apparent also in Chile where, since 1973, 'the transnational corporations have operated with total liberty' (Lahera 1981, 123). Finally, the extent to which the problems created by this pattern of accumulation based on comparative advantage, the free market and the ascendancy of transnationals in Latin America has concerned writers is seen in the wide-ranging discussion of recent years (Abalo 1981; Armstrong 1981a; Bastias 1980; Ferrer 1981; Fortin 1979; Lahera 1981; Martel 1978).

Industrialization and urban change

We turn now to the second part of this overview: to consider the interaction of such patterns of capital accumulation with urban development in Latin America. It seems redundant, perhaps, to observe that the maturing of capitalism in the continent has led to spatial structures and relationships comparable with those in the advanced industrial societies. The centralizing tendencies have distended the primate cities, causing them to overflow their limits, so

> leading to the formation of vast and complex regions . . . made up of a growing conurbation . . . with a network of smaller urban centres and rural areas directly linked to this nucleus: these regions become *central sub-systems* of the national system and act as the true motive centre of the latter.
>
> (de Mattos 1982, 72)

According to de Mattos, it is logical that the industrialization process in each country should have started in established major cities. They offered overwhelming advantages for the profitable absorption of investment capital; the widest access to the domestic market; an inherited infrastructure of transport, communications and public facilities; a concentrated pool of labour; cultural and recreational facilities for the personal needs of businessmen; and access to the seat of national political power and decision making.

Adopting a phrase of Topalov (1979, 20) that 'as a spatialized system of elements, the city is a form of capitalist socialization of the forces of production', which provides capital with '*useful effects of agglomeration*', de Mattos argues that the metropolitan centres were the natural choice for the location of transnational activity. In turn, the centralizing/concentrating capacities of the monopoly corporations have reinforced existing cumulative tendencies and speeded the transition from primate city to central sub-system: 'economic concentration and territorial concentration have tended to strengthen each other mutually' (de Mattos 1982, 73).

National state policies have pushed spatial concentration still further. The reasons for this are evident: the central sub-systems act as the theatres of action for ruling alliances who exercise a strong and continuing influence on state policies; they are also the place of residence of large concentrations of population with considerable, often disproportionate, political weight close at hand; and, finally, the very size of established economic activity concentrated in the central sub-system demands a constant flow of new investment on infrastructure and services to maintain its perceived efficiency and dynamism (de Mattos 1982, 75).

Yet, having stressed the characteristics of industrial capitalist growth common to large urban centres, we need to distinguish among particular aspects of urban expansion in different Latin America societies. We start by examining Castells' description of Latin America's major cities (or central sub-systems) in the phase of state-corporate modernization:

> an urban population unrelated to the productive level of the system; an absence of a direct relation between industrial employment and urbanization, but a

> link between industrial production and economic growth; a strong imbalance in the urban network in favour of one predominant urban area; increasing acceleration of the process of urbanization; a lack of jobs and services for the new urban masses and, consequently, reinforcement of the ecological segregation of the social classes and a polarization of the system of stratification as far as consumption is concerned. (Castells 1977, 57)

In the following pages we have constructed a number of Tables (4.1 to 4.7) which set out basic demographic and economic information for selected countries, and so help to provide a framework for the discussion.

Table 4.1 shows that population continues its rapid growth in most societies, although there is evidence of a recent slow-down in some. There are also great differences, from the low rates of Argentina and Uruguay at one end of the scale, to Honduras, Mexico and Venezuela at the other. The employment consequences of high growth rates are underlined by the Latin American Human Settlements Conference: by the year 2000, Latin America's potential workforce will increase from eighty million to more than 200 million. In addition, the movement of people from the countryside into the cities will, by the end of the century, bring over one-third of the population to live in thirty-seven cities of more than a million inhabitants (*Comercio Exterior* 1979a, 1230).

Latin America is now a predominantly urban continent. Argentina, Chile and Uruguay have been joined recently by Brazil, Colombia, Mexico, Peru and Venezuela, the most populous countries in the region, all with more than two-thirds of their people in urban settlements (taking 2000 as the threshold for an urban centre). This trend will continue to the end of the century; 75 per cent of the 625 million people will be urban, and the rapid pace of urbanization will represent an addition of 10 million people to the continent's cities each year (Gutman 1982, 1305).

Rural-urban migration has been a dynamic demographic feature of recent decades. In the 1960s, 72 per cent of São Paulo's increase, 54 per cent of that of Caracas, and 33 per cent of Bogota's growth were provided by such immigrants (Gutman 1982, 1305). When combined with high natural growth rates, this has created an urban explosion (Tables 4.1 and 4.2), especially in Brazil, Colombia and Peru, as well as in the Central American countries where rapid urbanization rates stem, in part, from low base figures. The traditionally urban societies, Argentina, Chile and Uruguay, have had much slower growth although they remain the most highly urbanized in Latin America.

One other feature of this growth stands out: in most cases, the primate city became more important in the decade 1960–70. Buenos Aires, Montevideo and Santiago are the most extreme examples of centralization, but the trend is similar in most other countries. It is also probable that demographic concentration will

Table 4.1 Latin America: patterns of urbanization, 1960–80

	Population 1980	*Average annual population growth rate*		*Urban population as % of total population*		*Average annual urban population growth rate*		*Average annual rural population growth rate*		*Population of largest city as % of total population*		*Average annual urban population growth rate 1960–80*	*Average annual rural population growth rate 1960–80*
		1960–70	*1970–80*	*1960*	*1980*	*1960–70*	*1970–80*	*1960–70*	*1970–80*	*1960*	*1970*		
	000,000s	%	%	%	%	%	%	%	%	%	%	%	%
Argentina	27.06	1.5	1.3	73.6	82.7	1.97	1.84	–0.75	0.88	34.0	36.1	1.91	–0.82
Bolivia	5.72	2.6	2.7	29.9	41.5	4.77	3.81	1.54	1.78	12.3	14.5	4.30	1.66
Brazil	126.37	2.9	2.9	40.2	64.1	6.51	4.31	0.00	0.76	4.5[1]	4.6[1]	5.40	0.37
Chile	11.10	2.1	1.7	63.8	80.8	3.69	2.59	–1.81	–0.72	25.9	30.7	3.14	–1.26
Colombia	26.90	3.1	2.4	47.0	67.3	5.76	4.54	0.69	0.91	9.7	12.8	5.15	0.80
Dominican Republic	5.94	3.4	2.8	28.8	47.1	6.63	5.23	1.68	1.98	12.1	16.8	5.92	1.83
Ecuador	8.02	3.0	3.1	36.1	45.0	4.32	4.59	2.82	2.28	11.4	12.6	4.45	2.55
Guatemala	7.26	3.1	3.1	27.3	38.6	5.12	4.25	1.93	2.12	13.4	13.5	4.68	2.03
Honduras	3.69	3.2	3.4	20.8	40.2	8.25	5.43	1.52	2.37	7.1	10.2	6.84	1.95
Mexico	69.99	3.3	3.4	54.0	69.1	4.29	4.60	0.49	1.82	15.3	20.8	5.52	1.16
Nicaragua	2.74	3.0	3.3	33.8	54.3	6.21	4.82	–0.78	1.02	15.4	21.7	5.78	0.12
Peru	17.77	2.9	2.8	38.7	65.4	7.32	4.24	0.29	0.12	10.5	9.7	4.23	0.21
Uruguay	2.95	1.1	0.4	74.9	86.6	2.60	1.43	–2.57	–1.23	44.7	44.5	2.01	–1.90
Venezuela	14.91	3.5	3.4	62.9	78.8	4.56	3.90	2.24	1.29	8.1	6.0	4.45	1.77

Notes: [1] 1960 Rio de Janeiro; 1980 São Paulo.

Urban = centres of more than 2000.

Sources for Tables 4.1 to 4.7: FAO Production Yearbook, 1980, Table 3, pp. 64–6; *Statistical Yearbook for Latin America, 1979,* Table A1, p. 3; *Statistical Yearbook for Latin America, 1973,* vols I and II, pp. 64, 102, 134, 170, 244, 282, 342, 400, 464, 554, 586, 614, 642 (for 1960 statistics); *Statistical Yearbook for Latin America, 1976,* Projections, pp. 62, 97, 124, 156, 232, 268, 328, 382, 442, 526, 558, 584, 612; *Statistical Yearbook for Latin America, 1979,* Table 1A5, p. 6; *World Development Report, 1981,* Table 19, pp. 170–1; Table 3, pp. 138–9, except Guatemala for which see J. W. Wilber and S. Haber (eds) *Statistical Abstract of Latin America,* vol. 21 (1981), Table 2214, p. 286, Table 9, pp. 150–1, Table 10, pp. 152–3.

Table 4.2 Latin America: rural-urban patterns, 1960–80

	Population[1]						*Population increase/decrease*					
	1960		*1970*		*1980*		*Urban*		*Rural*		*Urban*	*Rural*
	Urban 000s	*Rural 000s*	*Urban 000s*	*Rural 000s*	*Urban 000s*	*Rural 000s*	*1960–70 %*	*1970–80 %*	*1960–70 %*	*1970–80 %*	*1960–80 %*	*1960–80 %*
Argentina	15,341	5,509	18,637	5,111	22,384	4,680	21.48	20.10	–7.22	–8.43	45.91	–7.71
Bolivia	1,104	2,592	1,760	3,020	2,560	3,602	59.42	45.45	16.51	19.27	131.88	38.96
Brazil	28,292	42,035	53,166	42,038	81,052	45,337	87.91	52.45	0.00	7.85	186.48	7.85
Chile	4,902	2,781	7,044	2,325	9,095	2,165	43.69	29.12	–16.40	–6.88	85.53	–22.66
Colombia	7,458	8,419	13,056	9,019	20,345	9,870	75.06	55.82	7.13	9.44	172.79	17.23
Dominican Republic	901	2,228	1,712	2,632	2,850	3,203	90.01	66.47	18.13	21.69	216.32	43.76
Ecuador	1,561	2,762	2,382	3,649	3,732	4,571	52.59	56.68	32.11[2]	25.27	139.08	65.60
Guatemala	1,082	2,883	1,782	3,490	2,701	4,305	64.70	51.75	21.05	23.35	149.63	49.32
Honduras	385	1,464	851	1,702	1,444	2,151	121.03	69.68	16.26	26.38	275.06	46.92
Mexico	19,467	16,579	29,615	20,698	46,432	23,533	52.13	56.79	24.84	13.70	138.52	41.94
Nicaragua	507	994	926	1,044	1,483	1,250	82.64	60.15	5.03	19.73	192.50	25.75
Peru	3,877	6,147	7,858	5,688	11,901	6,296	102.68	51.45	–7.47	10.69	206.96	2.42
Uruguay	1,904	638	2,460	495	2,834	438	29.20	15.20	–22.40	–11.52	48.84	–31.35
Venezuela	4,869	2,872	7,603	2,956	11,142	2,992	56.15	46.55	2.92	1.22	128.84	4.18

Notes: [1] Information from *Statistical Yearbook for Latin America*, 1973, 1976; [2] 1976 CEPAL projections.

continue. From 1975 to the year 2000, cities or central sub-systems of more than five millions will increase 3.5 times; cities from two to five millions by three times; from 500,000 to two millions by 2.3 times; and those from 200,000 to 500,000 by 1.7 times. To meet housing needs satisfactorily, such growth would require building a new Buenos Aires each year (Gutman 1982, 1305).

Demographic weight is just one indicator of the concentrated influence of the metropolitan centres. Mexico City, Latin America's premier example of a central sub-system, illustrates the wider impact: occupying 2400 km^2 or 1.2 per cent of the national area, it contains rather more than 20 per cent of the nation's population. But it employs 43 per cent of the industrial workers, 42 per cent of those in commerce, absorbs 43 and 51 per cent respectively of the capital invested in industry and commerce, generates 43 per cent of the value of industrial production, and a similar proportion of net sales in commerce.

> It is the political and economic centre of the country to which more than a thousand people migrate daily in search of a small crumb of progress. It is the place where the splendour of prosperity and the misery of underdevelopment live together. (*Comercio Exterior*, 1979b, 1084)

Urbanization still continues outside the major centres, however. Given the rapid rates of urban growth which appear in Tables 4.1 and 4.2, a significant part of urban expansion takes place in regional cities, though more research is needed to learn about the roles being played by cities of different sizes in the continent (Roberts 1978, 86–7). We will take up this point later.

Turning to the rural areas, we can see from the tables that, despite heavy flows to the cities, rural population has continued to grow in most countries. Only in Argentina, Chile and Uruguay has there been a decline. In general, increases have been low, at less than 2 per cent annually, but the rates of rural growth were higher in the period 1970–80 than in the previous decade (Tables 4.1 and 4.2). This coincides with declines in urban growth rates in the same periods. The rural population is by no means a negligible factor, economically, socially or politically, especially in the less industrialized societies.

Table 4.3, urbanization indicators 1960–80, summarizes some of the important urban, rural and national demographic trends of the past 20 years. It also illustrates the diversity of national situations in the continent and allows us to distinguish three broadly defined groups of countries. The first comprises Argentina, Chile and Uruguay which share the characteristics of low population growth rates, high urbanization levels (joined here by Venezuela), low urbanization growth rates (medium in Chile), high primate city concentration levels and an absolutely declining rural population. This is the most homogeneous group.

Table 4.3 Latin America: urbanization indicators, 1960–1980

	National population growth rates			*Urbanization levels (1980)*			*Urbanization growth rates*			*Primate concentration*			*Rural population growth*		
	High	*Medium (2–3%)*	*Low*	*High*	*Medium (50–70%)*	*Low*	*High*	*Medium (3–5%)*	*Low*	*High*	*Medium*	*Low*	*Growth (0–1%)*	*(>1%)*	*Decline*
Argentina			X	X					X	X					X
Bolivia		X				X		X				X		X	
Brazil		X			X		X					X	X		
Chile			X	X				X		X					X
Colombia		X			X		X					X	X		
Dominican Republic	X					X	X				X			X	
Ecuador	X					X		X				X		X	
Guatemala	X					X		X				X		X	
Honduras	X					X	X					X		X	
Mexico	X				X			X			X			X	
Nicaragua	X				X		X				X			X	
Peru		X			X		X				X		X		
Uruguay			X	X					X	X				X	
Venezuela	X			X				X				X	X		

Table 4.4 Latin America: employment characteristics, 1960–79

	Proportion of population of working age (15–64)		*Proportion of labour force[1] in:*					
			Agriculture		*Industry*		*Services*	
	1960 %	*1979* %	*1960* %	*1979* %	*1960* %	*1979* %	*1960* %	*1979* %
Argentina	64	63	20	13	36	28	44	59
Bolivia	55	53	61	50	18	24	21	26
Brazil	54	55	52	40	15	22	33	38
Chile	57	62	30	20	20	20	50	60
Colombia	50	59	51	27	19	21	30	52
Dominican Republic	49	52	67	50	12	18	21	32
Ecuador	52	52	57	52	19	18	24	30
Guatemala	51	53	67	56	14	21	19	23
Honduras	52	50	70	63	11	14	19	23
Mexico	51	51	55	37	20	26	25	37
Nicaragua	50	50	62	40	16	14	22	46
Peru	52	54	53	38	19	20	28	42
Uruguay	64	63	21	11	29	32	50	57
Venezuela	51	55	35	19	22	27	43	54

Note: [1] Includes economically active population (plus armed forces and unemployed); excludes housewives, students and economically inactive groups.

Brazil, Colombia, Mexico, Peru and Venezuela form the next group. Each has medium levels of urbanization (somewhat arbitrarily set at between 50 and 70 per cent of the total population), medium (3 to 5 per cent per year) urban growth rates (except for Colombia and Peru) and the weight of the primate city in total population remains quite low (medium in Mexico and Peru). Rural population continues to grow slowly, except in Mexico. National population growth rates are medium level, though high in Mexico and Venezuela.

The third group consists of Bolivia, the Dominican Republic, Ecuador, Guatemala, Honduras and Nicaragua. Here we find generally high national population growth rates (medium in Bolivia), low urbanization levels (medium in Nicaragua), medium to high urban growth rates, low urban concentration levels in the major city (medium in Nicaragua and the Dominican Republic) and higher rates of rural population increase than in the other groups.

Each group is representative of particular historical phases of social and economic development. Argentina, Chile and Uruguay, together with Brazil

Table 4.5 Latin America: distribution of GDP, 1960–79

	Agriculture[1]		*Industry*[2]		*Manu-facturing*[3]		*Services*	
	1960	*1979*	*1960*	*1979*	*1960*	*1979*	*1960*	*1979*
Argentina	16	13	38	46	32	37	46	41
Bolivia	26	17	25	29	15	13	49	54
Brazil	16	11	35	38	26	28	49	51
Chile	11	8[4]	38	37[4]	23	24[4]	51	55[4]
Colombia	34	29	26	28	17	21	40	43
Dominican Republic	27	19[4]	23	26[4]	17	16[4]	50	55[4]
Ecuador	33	15	19	37	14	19	48	48
Guatemala[5]	33	30	15	18	12	14	53	52
Honduras	37	32	19	26	13	17	44	42
Mexico	16	10	29	38	23	29	55	52
Nicaragua	24	29	21	28	16	24	55	43
Peru	18	10	33	43	24	26	49	47
Uruguay	19	13	28	37	21	31	53	50
Venezuela	6	6	22	47	n.a.	16	72	47

Notes: [1] 'Agriculture' includes forestry, hunting and fishing; [2] 'industry' includes manufacturing, mining, construction, electricity, water and gas; [3] 'manufacturing' is part of industry; [4] 1978; [5] all figures at current price series except for figures where 1960 and 1979 are at constant 1970 dollars.

and Mexico, have experienced various forms of modernized industrial growth stretching back to the turn of this century. Colombia, Peru and Venezuela have a more recent postwar history of such activity, while the rest are newcomers to modern industrial capitalist growth. The nature of the industrial sectors thus varies considerably, some having been founded through national development strategies early in this century, while others have emerged more recently in association with foreign capital. The impact of the current wave of transnational penetration must therefore be understood in terms of these different historical experiences.

Some of the economic differences can be seen in Tables 4.4 and 4.5, the former highlighting variations in sectoral employment. Argentina, Chile, Uruguay and Venezuela had less than 20 per cent of their workforces in agriculture in 1979, Brazil, Colombia, Mexico, Nicaragua and Peru had between 20 and 40 per cent, and in the rest more than 50 per cent. The industrial scene is less clear, in part because the differences in industrial employment do not distinguish between modern capitalist and labour-intensive craft activities, although this is clarified in Table 4.5. Here the proportion of domestic product derived from 'manufacturing'

Table 4.6 Latin America: distribution of primary commodities and manufacturing (%),[1] 1960–78

	Fuels, minerals, metals		*Other primary commodities*[2]		*Textiles and clothing*		*Machinery transport equipment*		*Other manufactures*		*Total manufactures*	
	1960 %	*1978* %	*1960* %	*1978* %	*1960* %	*1978* %	*1960* %	*1978* %	*1960* %	*1978* %	*1960* %	*1978* %
Argentina	1	2	95	72	0	3	(–)	8	4	15	4	26
Bolivia	n.a.	88	n.a.	10	n.a.	1	n.a.	(–)	n.a.	1	n.a.	2
Brazil	8	11	89	55	0	4	(–)	15	3	15	3	34
Chile	92	74	4	21	0	(–)	0	(–)	4	15	4	15
Colombia	19	5	79	78	0	5	(–)	2	2	10	2	17
Dominican Republic	6	4	92	75	0	(–)	0	1	2	20	2	21
Ecuador	0	41	99	57	0	1	0	0	1	1	1	2
Guatemala	2	1	95	78	1	5	0	1	2	15	3	21
Honduras	5	(–)	93	90	0	1	0	0	2	9	2	10
Mexico	24	39	64	31	4	3	1	10	7	17	12	30
Nicaragua	3	1	95	82	0	2	0	1	2	14	2	17
Peru	49	46	50	43	0	3	0	1	1	7	1	11
Uruguay	n.a.	1	71	56	21	20	n.a.	3	8	20	29	46
Venezuela	74	97	26	1	0	(–)	0	(–)	(–)	2	(–)	2

Notes: [1] (–) = less than 0.5%; [2] includes food and live animals, beverages and tobacco, inedible crude materials, oils, fats, waxes.

alone shows Argentina, Brazil, Mexico and Uruguay to be the most highly industrialized in 1979, with Chile, Nicaragua and Peru some distance back.

The importance of the service sector is also evident. In most cases it is the largest contributor to domestic product (a feature we noted in Chapter 3) and is a major factor underlying the growth of the continent's large cities and subsystems. The urban areas provide the conditions for both capitalist and non-capitalist commercial and service activity, a situation which, in part, parallels that of the advanced capitalist countries where there has been a sectoral displacement from capital-intensive manufacturing production into urban service sector activities.

Table 4.6 widens our knowledge about the character of manufacturing activity. Argentina, Brazil, Mexico and Uruguay are still clearly predominant in export manufacturing (see also Movarec 1982, 47–77), but the transformation of the export sectors of other economies is the striking feature of the period 1960–78. Growth is evident everywhere, with the highest rates in Brazil, Colombia, the Dominican Republic, Guatemala, Nicaragua and Peru. (Although we cannot elaborate further here, it is observable that agro-processing is an important element in the expansion of some economies.) It is notable that FTZs were established in four of these societies in the 1960s and 1970s (Fröbel, Heinrichs and Kreye 1980, 309), and are also important in Mexico, one of the largest manufacturing exporters in Latin America.

Table 4.7 on merchandise imports has been included to round out the information on the differential effects of industrialization – in rural as well as urban areas – on the fourteen selected countries. Food imports rose in a number of countries where agribusiness and local commercial farmers have been most active in modernizing the sector (Argentina, Colombia, Guatemala and Mexico). Fuel imports also grew dramatically, a consequence of rapid world price increases, and of the petroleum-dependent nature of the modernization process. Little can be said of the manufactured commodity imports in this brief review, other than to note that in most cases, little change has taken place if the 'machinery' and 'other' categories are aggregated. A closer inspection would reveal interesting patterns and changes if the transformation ('machinery and transport') and 'other' categories were to be disaggregated.

We should sound a note of caution about the FTZ industrialization strategy; while significant in recent export-oriented development, it is still far from playing the central role that it does in certain Asian economies. Industrial expansion in Latin America has a longer and more complex history. First, a country such as Argentina, without FTZs (although with an export-slanted free market policy since 1976) had higher rates of manufactured export growth than Mexico with its FTZs, in the period 1960 to 1978.

Table 4.7 Latin America: import distribution (%), 1960–78

	Food		*Fuels*		*Other primary commodities*		*Machinery, transport equipment*		*Other manufactures*		*Total manufactures*	
	1960 %	*1978* %	*1960* %	*1978* %	*1960* %	*1978* %	*1960* %	*1978* %	*1960* %	*1978* %	*1960* %	*1978* %
Argentina	3	6	13	12	11	9	44	39	29	34	73	73
Bolivia	n.a.	n.a.	n.a.	n.a.	n.a.	n.a.	n.a.	n.a.	n.a.	n.a.	n.a.	n.a.
Brazil	14	10	19	33	13	6	36	26	18	25	54	51
Chile	n.a.	15	n.a.	16	n.a.	13	n.a.	22	n.a.	34	n.a.	56
Colombia	8	11	3	7	15	7	43	36	31	39	74	75
Dominican Republic	n.a.	17	n.a.	22	n.a.	4	n.a.	23	n.a.	34	n.a.	57
Ecuador	13	7[1]	3	1[1]	9	4[1]	33	50[1]	42	38[1]	75	88[1]
Guatemala	12	15	10	16	7	39	26	23	45	7	71	30
Honduras	13	9[1]	9	12[1]	3	2[1]	24	31[1]	51	46[1]	75	77[1]
Mexico	4	13[1]	2	3[1]	10	7[1]	52	45[1]	32	32[1]	84	77[1]
Nicaragua	9	10	10	15	5	2	22	24	54	49	76	73
Peru	16	16[1]	5	19[1]	5	4[1]	37	33[1]	37	28[1]	74	61[1]
Uruguay	5	7	24	32	46	9	17	25	8	27	25	52
Venezuela	18	12	1	1	10	4	36	52	35	31	71	83

Note: [1] 1977.

Second, the larger nations, with their more advanced and complex manufacturing sectors,

> tended to increase their industrial weight within the region due to the expansion of Brazil, and, to a lesser extent, Mexico, and despite the fact that Argentina, which sharply increased its degree of industrialization, did so within a context of only slow growth. (Soza 1981, 42)

And finally, while similarities to the Southeast Asian experiences of FTZ expansion may be found in the Dominican Republic, Guatemala and El Salvador (before the recent ferment), the larger economies, with already diversified industrial structures, are closer to the South Korean model, though probably lacking the same degree of national autonomy over the capitalist sector. Calcagno and Jacobowicz (1981, 25–6) note that in Brazil and Mexico less than 9 per cent of total product is exported, and less than 5 per cent of manufactured production goes abroad. Both countries 'are typical examples of industrialization by import substitution, with the complementary export of a minor part of industrial production'. They point out, too, that the FTZ *maquila* activity accounts for only 2.3 per cent of Mexican industrial production. Much of the manufacturing sector, whether TNC or national, is thus producing for the domestic consumer market.

There are certain key features to this modern industrialization. First, it is highly imitative of foreign patterns. Whether we consider the transnational branch plant or the national firm, units are large-scale, highly concentrated and capital-intensive. The transnationals remain the pacesetters in the modernization strategy,[3] providing much of the imported management and technology. The capital intensity even in plants established to take advantage of low labour costs shows up in the comparison between the rapid growth of industry and the much slower growth of manufacturing employment.

A recent study notes that only in the initial phases of industrialization in Latin America has manufacturing increased its share of employment in any economy. Later, the declining capacity of modern production to absorb workers has forced the workers to seek jobs elsewhere, usually at low pay and productivity levels (UNESCO/CEPAL/PNUD 1981, 149). Tavares and Teixeira also observe that national enterprise must follow a path of 'defensive modernization' to compete with foreign technology; the model is thus weighted inherently against an equivalent growth in employment (Tavares and Teixeira 1981, 95; also see Girvan 1976, 7).

Intra-urban circuits, modernization and consumption

The consequence of all this, Milton Santos has argued, is to create a situation in which the Latin American city

> can no longer be studied as a single homogeneous entity, but should be thought of as two sub-systems, namely the 'upper' or 'modern circuit' and the 'lower circuit'. (Santos 1979, 8)

The upper circuit is geared heavily to export and re-export activities as well as to a restricted upper income market for local consumer goods. It is, in general, he suggests, indifferent to the needs of local popular consumption, for such production contributes little to forming further linkages within the national economy, or creating autocentric development.[4] The multiplier effects are mostly enjoyed by other economies. The spatial consequences of this can be seen in the growth of specialist economic activity in locations 'integrated to foreign production and distribution circuits' (Santos 1979, 183).

The lower circuit, however, 'plays an undeniable social and economic role, for it not only absorbs "surplus labour" but most of the people who are expelled from the countryside' (Santos 1979, 203). Even so, it also perpetuates poverty by collecting the daily savings and earnings of low-income groups which it transfers to the capitalist sector through financial and/or political channels.[5] This parasitic relationship which constitutes the principal link between the two circuits, will need to be replaced by a spatial and productive reorganization which allows an improvement in the productivity and life-condition of those in the lower circuit (Santos 1979, 205ff.).

Any attempt to reorient policy, however, will run into heavy opposition. There would be severe confrontations with the system's beneficiaries, the corporate and political élites especially, and the state strategies (internal and external) would need to be radically restructured. But, as we noted in Chapter 3, the levels of resistance probably run deeply through more layers in Latin America societies than Santos allows.

Producers of modern industrial goods have usually aimed first at serving a limited high-income local clientele; expansion of modern sector production has thus been traditionally directed to 'exploiting the upper segment of the market more intensively rather than through extending the market' (Jenkins 1979, 178). Corporate interest, in the past especially, has aimed to promote growth in demand by high income groups, and so the bias in modernized development strategies has been to maintain a structured inequality in income distribution. Both in terms of production (low costs, especially for labour) and of consumption (stimulated élite demand) the pressures toward inequality have been deeply embedded in the dependent urban industrialization model.

Hymer argues that capitalism could have followed either of two paths. One was to widen the production process to make consumer goods generally available throughout the world; the other was to provide a constant stream of new consumer innovations for a more limited market.[6] The capital deepening resulting from the second alternative had a double effect, for it also slowed the rate of growth of the industrial labour force. The production structures and, in fact, the entire accumulation process, were thus reinforced by imitative consumer styles among the urban upper income groups of the continent.

Nevertheless the case argued by Santos and Hymer in the 1970s now needs amendment, for modernization in recent years has altered the consumption habits of Third World populations. The effects are now extended beyond the wealthy few, with the penetration of derivative consumer behaviour into other classes and regions outside the main centres.

Modern factory products circulate so widely that even rural areas and market towns now display them alongside those of local craftspeople. Imported or locally produced clothing, cooking ware and utensils and simple tools appear, as we shall see later, in the stalls of remote Ecuadorian villages. Together with other factory products such as transistor radios, they are sought after to the extent permitted by income, so that consumer reliance upon the modern sector is coming to permeate all levels of society.[7]

The issues facing Latin American societies are thus wider than those of commercial dependence or the domination of capitalist production: the diffusion of modern lifestyles and the manipulation of consumer demand further deepen the existing technological and financial dependence. The urban upper and middle classes are most affected by these trends, but other groups are also increasingly incorporated into capitalist lifestyles and consumer habits (Filgueira 1981).

Primate cities and central sub-systems play a crucial role in these social and economic changes. They are the home base of the collaborative ruling class which imports the ideologies, the behaviour patterns (in particular the consumer habits) and the products of the corporate sector. In turn the élite groups diffuse their tastes to the rest of society so that, as Filgueira (1981) has shown, basic consumer needs of the poor are reshaped, and limited family resources are spent on modern goods. Regional cities, market towns and rural areas, in turn, are persuaded to buy the goods produced in the major cities or imported from overseas. Thus disparities widen as the regions become increasingly dependent upon metropolitan or foreign products and, at the same time continue to provide surplus for capitalist sector growth in the major centres.

Capital accumulation, then, remains the prime function of the Latin American metropolis or central sub-system and its ruling classes. The diffusion of modern ideologies and habits supports and reinforces the accumulation drive of the capitalist sector by creating new markets and promoting the 'felt needs' for corporate consumer goods. Cardoso has argued that the city is typically the only area where the social class relations necessary for capitalist production are fully in existence (Cardoso 1975, 159); now, with corporate advertising and consumer promotion, the nexus between urbanization and capital accumulation becomes still stronger. As the process continues, the differences in development trajectories among Latin American societies may be narrowing (Castells 1977, 59ff.).

Convergence in Latin America's cities?

Roberts' argument (1978, 81ff.), that an urban convergence is taking place in the continent, is important here. The concentrations of high- and middle-income consumers, of large masses of population and supplies of labour, together with infrastructural services and facilities, all draw industry into the larger urban centres. And the more complex the urban environment, the tighter its involvement with overseas technology, know-how and financing in both production and consumption.

The obverse of this increasing metropolitan concentration, according to Roberts, is the relative decline of smaller centres: a growing regional imbalance within nations. This is true in many cases; yet the industrialization-urbanization relationship seems rather more intricate: Roberts himself seems ambiguous on the issue, as is apparent in his comment on technological dependence that:

> one of its most evident consequences is the stagnation of provincial regions in face of the concentration of economic growth in a few large cities.
>
> (Roberts 1978, 86)

Yet earlier, noting our lack of detailed knowledge about provincial economic and political activities in Latin America, he warns:

> It would be unwise . . . to assume that provincial regions are stagnant economically and unimportant elements in national decision making.
>
> (Roberts 1978, 86)

One resolution of this apparent ambiguity might be to accept that both of Roberts' statements can be correct, depending upon the societies and cities we are considering. To help distinguish among them, we can draw upon our earlier analysis based on the economic and demographic information in Tables 4.1 to 4.7. The historical changes in socio-economic development portrayed here also allow us to infer certain features of the spatial patterns of settlement and activity in each society. The current outcome of this historical evolution was summarized in Table 4.3 and the accompanying discussion of the three broad groups of societies according to their 'urbanization indicators'.

Such a typology is approximate, but it allows us to gain a deeper understanding of the diversity of Latin American experience. Those societies which entered into close relationships with the international economy early on are most likely to have concentrated demographic and economic growth around a major centre or sub-system with a concomitant dilution of secondary centre influence. The degree of concentration seems to be less (though varying) among societies entering the international system more recently.

Because the latter have been subjected to centralizing pressures of modernized growth for a shorter period, their secondary centres have probably enjoyed a greater latitude for independent expansion as regional market, service and even light industrial centres. Colombia's Cali and Medellin are prime examples of regional cities which have rivalled Bogota, until recent years when capitalist development has promoted the growth of the national capital to a greater extent. Similarly, special circumstances have favoured regional growth poles, as in the historical examples of plantation or mining enclaves such as Huancayo, Recife or Salta and modern state-created centres such as Ciudad Guyana or, indeed, Brasilia. Under different historical conditions, then, Roberts' second proposition is clearly applicable.

The strategy of creating Free Trade Zones to attract foreign capital has further promoted regional growth in cities such as Manaus in Brazil, Barranquilla and Buenaventura in Colombia, and, above all, along the frontier zone in the north of Mexico (Tijuana nearly tripled its population between 1960 and 1977 (Sargent 1982, 227).

Foreign investors, too, have discovered locational advantages in depressed hinterland regions where costs are lower. Santos notes that modern technology, state subsidies and export-promotion policies allow corporations to establish in a widening variety of locations, a situation which is attractive for those not concerned with the internal market.[8] This gives another twist to the concept of 'footlooseness'. Not only can firms pick up and leave for more profitable, lower cost locations, but they do not need to adapt to the classical economics of localization and urbanization which draw the medium-sized factory, serving the local market, to the larger cities. Production for overseas markets, regional subsidies for infrastructure and depressed wage costs all provide incentives for decentralization.

Over time, therefore, the patterns of urbanization have varied among the different societies. The historical interaction between internal and external forces led to a variety of national paths; some societies (and regions within them) developed modernized social and economic structures, and this, in turn, was reflected in advanced urban development. By the early twentieth century, for instance, the societies of settlement (see Chapter 2) had become incorporated into the trade-investment-migration mechanisms of a rapidly expanding international economy. In the southern cone countries especially, the internal response of urban-based ruling coalitions was to use the power of the nation state to collaborate and promote their own interests (Cardoso and Faletto 1969–78).

Buenos Aires, Montevideo, Rio de Janeiro, São Paulo and Santiago then became the centres of operations for the strategy 'through which the economy and society became integrated in a subordinate manner into the international system' (Cotler 1979, 256). They provided the urban base from which the ruling alliances worked to achieve a degree of relative autonomy in their dealings with foreign interests. The resultant pattern of national capital accumulation in turn reinforced urban expansion, so creating modern financial, commercial and industrial cities from the early decades of this century. Such capitalist evolution was also reflected in the rise of an urban working class with an organized capacity to battle for its own class interests, so turning these cities into dynamic centres of social activity and change.

In the societies of enclave exploitation, however, the possibilities for local capital accumulation and internalized growth were more restricted; lacking the multiplier effects of integrated growth, urban development, too, was limited. In the absence of a strong entrepreneurial urban middle class, such societies tended to rely on foreign capital and expertise in their post-Second World War industrialization programmes so that diversification policies in countries such as Peru and Ecuador

> coincided with a period in which the ties of dependence were becoming stronger rather than weaker, and, for that reason, the substitution process was relatively independent of national capital. (Quijano 1975, 131)

Despite these varying histories, the past two decades of capitalist modernization appear to be leading to a gradual convergence in urbanization/industrialization patterns among the large metropolitan centres of the continent. In this, corporate expansion is strongly supported by state policies (de Mattos 1982), so that in those societies where popular opposition has been articulate and the labour movements are organized, bureaucratic-authoritarian regimes have enforced their modernization strategies through political repression to ensure the free operation of market forces (Collier 1979).

Under such conditions, and in those societies where workers are less organized, the continent's metropolitan centres have come to be regarded by corporate enterprise (both transnational and national) as theatres of operations in which capital accumulation can take place with minimal social or political hindrance.[9] Modernization strategies are creating a pattern of convergence in which the major cities are characterized by their commercial and financial influence as well as, more recently, their growing industrial activity. They are, further, centres of political and administrative decision making in which state-transnational-local oligopoly collusion is designed to ensure the success of this growth model.

Convergence in consumer patterns is also occurring in the principal urban areas. The ruling groups in the primate cities and central sub-systems provide the primary markets for imported products and lifestyles, and act as conduits to the rest of society; similarly the national metropolises become the centres producing or importing Hymer's 'continuous innovations'. We can thus observe inter- and intra-national diffusion occurring; whether the expenditure on many modern consumer products serves the best interests of those whose basic needs remain unsatisfied, or of small-scale producers under the challenge of mass-produced industrial goods, is another question.

But there is also evidence of a counter tendency: divergence. While the social and spatial diffusion of consumption patterns is occurring, we can question whether there is an equivalent trickle-down of modern productive activity. Apart from examples of decentralized FTZ export manufacturing activity in specially designated areas and other cities such as Cali, Guadalajara and Cordoba, most capitalist industrial activity continues to concentrate around the metropolises.

Growing disparities and dependence both between and within countries are the result. With the undermining of local manufacturing and farm production, the possibilities for any real degree of national self-sufficiency and independence in satisfying basic needs are diminished. In turn, the dependence of regional cities, market towns and rural areas on the output of central sub-systems grows as factories concentrate in the large centres. This converts secondary and tertiary centres in the urban hierarchy into distribution points for such products to village and rural markets further down the line.

These trends intensify with continuing modernization in the continent. Both the direct impact on non-capitalist production and distribution systems and the indirect undermining of 'lower circuit' activity by the spread of imitative consumption habits deepen social and economic pressures on the low income groups. Here, the primate cities, the national headquarters for local ruling groups and transnational corporations, play a crucial promotional role both as theatres of accumulation and entrepôts for the diffusion of imported lifestyles.

Confrontation and continentalization

Other issues arise from this survey. First, the social and regional disparities associated with concentrated growth will not be resolved merely by spatial relocation of economic activity. Regimes which are committed to free market politics are, in any case, likely to let the market decide location. But even where decentralization is tried, the results might be other than expected. Citing Venezuelan and Brazilian examples, Gilbert argues that 'establishing industrial activity in provincial cities will merely create at the provincial level the national problems of dependent development'. (1982, 272). We also take up this issue in Chapter 7.

Industrializing regional cities similarly reflect current patterns of Third World capital accumulation: dependent, oligopolistic, élite-serving and only marginally relevant to the basic employment and consumer needs of the majority (Coraggio 1972, Armstrong 1973). Some form of regional decentralization is required in the continent. But merely to change the geographic distribution of industrial activity may well be as counterproductive for low income groups as the recent corporate modernization of Latin America's rural economy has been for the rural population.

Second, while the continent's large cities are prime sites for corporations, they also furnish living space for increasing numbers of working-class and marginalized people. They are becoming the arenas for confrontation between capital and the state, and an increasingly restive, though still largely unorganized, urban majority.

The existence of discontented urban groups crowded into the principal cities poses a challenge to current economic strategies and systems of political control. Most workers are not yet strongly organized (with the exception of Argentina's urban workers who have survived harsh repression in recent years, and Brazil's unions which have re-emerged after two decades of military rule), yet there are 'powerful forces acting to homogenize their situation . . . [and] it makes the working class an evidently volatile element in urban politics' (Roberts 1978, 176). Urban disaffection has also thrown up leaders and organizations which are now mobilizing massive discontent, especially in the regions and rural areas of Central America.

In the societies of the southern cone, authoritarian military regimes have strained to keep the lid on urban unrest. But the weaknesses of the strategy – severe political repression, social and economic constraints on the working and lower-middle classes, growing external dependence, and the heavy burden of foreign indebtedness – are clearly calling into question the viability of the corporatist free market approach.

One final aspect of the accumulation-urbanization relationship requires mention. We have discussed the urban question largely in terms of its national setting. But with the internationalization of capital and its concentration into fewer but larger urban entities, the possibility of a continental hierarchy of cities and sub-systems emerging is already here.

If Hymer (1975) and Friedmann and Wolf (1982) are correct, then we can think of enormous central sub-systems such as Mexico City, São Paulo and, perhaps, Buenos Aires acting as the major decision centres in Latin America, and as secondary or regional metropolises in the global hierarchy (New York, Tokyo, Frankfurt or London being in the top rank). The wide acceptance of the need for free labour and capital markets, and the pressure of transnationals and international agencies for a truly Latin American Free Trade Area by the erosion of national barriers to the movement of capital and labour could make this a believable outcome.

The implications for intensifying social inequality and growing regional disparities on a continental scale, given the present trends toward concentration, are hard to avoid. Capital and labour movements which today occur freely within countries would be exacerbated by transfers between countries (such as, for example, Paraguayan migration into Brazil and Argentina); at the same time, social, economic, political and cultural tensions would be heightened as the already crumbling sovereignty of the smaller nations gave way to the logic of continental-scale capital accumulation.

It is not easy to predict the social and political outcome of these trends, but at least three scenarios suggest themselves: first, the spread of urban-led, peasant-based insurrections, as in the rural societies of Central America; second, a return to civilian nationalist regimes pursuing policies of national and more internally oriented and directed growth, as has occurred in the industrially advanced societies of Argentina and Brazil; and finally, a less likely possibility, at least in the foreseeable future, in which popular opposition assumes continent-wide proportions to confront the activities of capital operating freely throughout Latin America.

5

The urbanization process, accumulation and industrialization in the market economies of Asia

Introduction

The accumulation-urbanization process in the market economies[1] of the developing countries of Asia is characterized by a different mix of demographic, economic and social factors from that of Latin America. First, by the beginning of the 1980s the countries that made up this region (excluding the socialist countries of China, North Korea, Vietnam, Kampuchea and Laos) contained well over 1 billion people compared to the Latin American figure of 330 million for the fourteen major countries. Much of this population is, of course, concentrated in three countries – India, Pakistan and Indonesia – but there are also large populations in Bangladesh, the Philippines, Korea and Thailand. Historically, too, there have been several important differences from Latin America, primarily relating to the length of experience with international and national capital. The Asian market economies were incorporated into the world economy during the mercantilist era and have only really begun to change their role and relationship in the international system with the independence of the majority of these countries in the brief period since 1945.

For most Asian market economies (unlike Latin America) this protected period of mercantilist relationships left a distinctive heritage both in terms of urban hierarchies dominated by one large primate city and in terms of truncated class structures in which alien communities played significant roles in trade while political power remained with colonial élites and 'tamed' Asian traditional élites. The period since 1945 has seen, however, a rapid, if recent, expansion of international investment often in conjunction with national state capital in many countries of the region. The pace of expansion has propelled some Asian market economies into modern capitalist economic growth as well as the proletarianization of significant parts of the population.

Finally, it may be argued that the cultural diversity of the Asian region together with the differential effect of a variety of colonial regimes have left a much more diversified region with many languages, government systems and degrees of incorporation into the world economy. In short, it seems much more difficult to talk about Asia as a unit than about Latin America.

By 1950 the large populations, relative lateness of political independence and the predominantly mercantile relationships left a major impact upon the accumulation-urbanization-industrialization process in Asia. Virtually all countries were characterized by low urbanization with an urban hierarchy often dominated by one primate city (Manila, Jakarta, Bombay, Colombo), by limited industrialization, and the export of one or two unprocessed or semi-processed raw commodities (for example, teak, tea, tin, rubber) to the markets of the colonial metropolises. In most cases the countries were ruled by political élites who had gained their power through national campaigns of 'decolonization'. Their links to both national and international capital were distant and for much of the early period of independence many were concerned to grapple with 'custodial' rather than 'developmental' goals. Looking at the situation of most of the Asian countries in the mid-1950s it is hard to believe that there were great prospects for economic growth and urban transformation. Political instability and high rates of natural increase appeared to be stimulating city population growth, while economic growth was insufficient to create employment opportunities in their wage sectors. The result was a proliferation of people in low-income service sectors.

By the early 1970s, it was clear that a much more divergent pattern of urbanization was emerging with some countries experiencing a rapid growth of manufacturing.

In Table 5.1 basic data concerning population, urbanization and economic features of these countries according to the classification of urbanization trajectory is presented. As with all macro-data these statistics should be treated with caution, since the urban data reflects each country's definition and the labour force data for 1980 is in some cases estimated.[2] At least four main urban trajectories can be suggested which reflect various forms of development.[3] First, there are countries in which the urban transformation is proceeding very slowly. For the four largest countries in the group – India, Bangladesh, Indonesia and Pakistan – which have some 75 per cent of the Asian market economies' population, urbanization levels have increased very slowly over the last decade. This has meant that the countryside has absorbed a large proportion of the total population increase. In India, for instance, non-urban areas absorbed 78.6 million of

Table 5.1 Selected data on Asian market economies, 1960–82

	Population (mid-1982)	*Average annual population growth rate*		*Urban population as % of total population*		*Average annual urban growth rate*		*Proportion of labour force in*					
								Agriculture		*Industry*		*Services*	
	000,000s	*1960–70 %*	*1970–82 %*	*1960 %*	*1982 %*	*1960–70 %*	*1970–82 %*	*1960 %*	*1980 %*	*1960 %*	*1980 %*	*1960 %*	*1980 %*
Type I													
Hong Kong	5.2	2.5	2.4	89	91	3.1	2.4	8	3	52	57	40	40
Singapore	2.5	2.4	1.5	100	100	2.4	1.5	8	2	23	39	69	59
Taiwan	18.4	2.7	1.9	36	68	3.7	3.2	56	20	11	33	33	47
Korea Rep.	39.3	2.4	1.7	28	61	6.2	5.0	66	34	9	29	25	37
Type II													
Malaysia	14.5	2.9	2.5	25	30	3.5	3.4	63	50	12	16	25	34
Philippines	50.7	3.0	2.7	25	38	4.2	3.8	61	46	15	17	24	37
Type III													
Thailand	48.5	3.1	2.4	12	17	4.7	4.3	84	76	4	9	12	15
Sri Lanka	15.2	2.4	1.7	18	24	4.4	2.5	56	54	13	14	31	32
Type IV													
Bangladesh	92.9	2.9	2.6	5	12	4.7	6.0	87	74	3	11	10	15
Burma	34.9	2.2	2.2	19	28	3.7	3.9	68	67	11	10	21	23
India	717.0	2.3	2.3	18	24	3.5	3.9	73	71	11	13	16	16
Indonesia	152.6	2.2	2.3	15	22	3.7	4.5	75	58	8	12	17	30
Pakistan	87.1	2.4	3.0	18	29	4.5	4.3	61	57	18	20	21	23

Sources: Dillinger (1979); World Bank (1979 and 1984).

the 108.3 million total population increase in the period between 1961 and 1971. So, despite the fact that the population in urban areas grew quite rapidly from natural increase, rural-urban migration and urban reclassification, the actual volume of net rural-urban migration was only a trickle of the potential sending population. In these countries the urban transformation has hardly begun and this raises serious questions concerning the future of urban centres. Even assuming a fall in rates of natural increase, the size of cities in these countries is likely to be very large and there are severe problems of labour absorption both in the countryside and the cities.

A completely different urban trajectory is being followed by the city states of Hong Kong and Singapore and the states of Taiwan and South Korea. In the two city states the major transformation has been in structural shifts in the labour force from agriculture and the growth of industrial employment. In addition to their historically important service function as entrepôts and trading centres in the region they have now become major industrial exporters. In Taiwan and South Korea there has been very rapid urbanization over the last two decades bringing them both to a point where approximately 50 per cent of their populations live in urban areas. This has been accompanied by a radical transformation in the labour force with the growth of industrial activities particularly for export.

A third type of urbanization trajectory is exhibited strongly by Malaysia and less clearly by the Philippines. In both countries there have been strong efforts to industrialize over the last two decades as well as to improve rural areas, but Malaysia seems to have moved more successfully into a phase of export industrialization which is reflected both in the rapidly increasing urbanization levels and changing structure of the labour force. In the 1970s, the Philippines has done less well, as is reflected in the statistics in Table 5.1.

Finally there is a fourth type of urbanization trajectory, which may be described as incipient industrialization, in which agriculture remains important. On the basis of the data in the Table it might be argued that both Sri Lanka and Thailand fit more appropriately into the fourth category of slow urbanization with large rural populations. But both have embarked recently upon vigorous policies designed to increase export-oriented industrialization, and this seems likely to lead them along an urban trajectory similar to that of Malaysia.

Urbanization in Asia: some close-ups

These variations in the urbanization process are best shown by a more detailed discussion of five countries, illustrating the four main trajectories of urbanization. As Table 5.2 shows, there are very sharp differences between the five countries.

Table 5.2 Key indicators of the five study countries, 1960–80

	Socio-economic features										
	Total population (mid-1981)	*GDP per capita US$*		*Urban population as % of total population*		*Distribution of GDP*					
						Agriculture		*Manufacturing*[1]		*Services*	
		1972	*1980*	*1960*	*1980*	*1972*	*1980*	*1972*	*1980*	*1972*	*1980*
	000,000s			%	%	%	%	%	%	%	%
Hong Kong	5.1	1102	4072	89	100	1.6	1.0	27.0	27.6	71.4	71.4
Korea	38.7	303	1426	28	55	30.2	17.2	19.9	31.3	49.9	51.5
Malaysia	13.8	444	1644	25	49	29.1	23.6	20.8	24.0	50.1	52.4
Philippines	49.5	212	732	25	37	28.6	25.5	26.3	28.0	45.1	46.5
Indonesia	151.0	91	473	15	20	40.8	31.4	18.4	23.8	41.6	44.8

Note: [1] Includes mining, quarrying.

Sources: World Bank, *World Development Report 1982*; Asian Development Bank, *Key Indicators of Developing Member Countries of ADB*, April 1982.

HONG KONG

In this list Hong Kong is something of an anomaly. It is a colonial city and yet is one of the major financial and industrial centres of the world. Established in 1841 as a British enclave on the coast of China, it remained a sleepy entrepôt port until after the Japanese defeat in 1945. In the period since that date, its population has grown from some 600,000 to an estimated 5 million by 1980. Much of the early population growth was the result of the movement of refugees from China, which contributed significantly to the total of almost 2.5 million by 1950. Since that date, natural increase has been a major contributor to population growth and has created one of the most crowded urban environments in the world, with the majority of the population concentrated in the 25 km^2 that constitute the urban centres of Hong Kong and Kowloon. The population is largely living in high-rise public or private housing at very high household densities.

A second facet of the remarkable transition that Hong Kong has experienced is the creation of a major manufacturing base. Beginning with the considerable inflow of labour, capital and entrepreneurial skills in the late 1940s, Hong Kong has become one of the Third World's foremost export producers of clothing and textiles, although in recent years the plastic products and electronics industries have become more significant. Most exports are directed to five developed countries, which in 1977 took almost two-thirds of the products; this development has been associated with a dramatic increase in direct foreign investment in manufacturing.

A third facet of this rapid growth is Hong Kong's position as a financial centre. By 1977 over 100 foreign banks were operating in the colony, as well as a very large number of international insurance companies, making Hong Kong the world's third largest financial centre. The phenomenal growth of 'loans and advances abroad' given by Hong Kong-based banks, which rose from $HK204 million to $HK20,778 million between 1969 and 1979, is a measure of the city's international expansion.

A final aspect is the international role of Hong Kong as a major tourist centre. Between 1958 and 1977 the number of tourist arrivals rose from 103,006 to 1,755,669. This represented some 48,100 tourists a day spending approximately $HK2,073 during each visit in 1977, and generating some 100,000 jobs as well as adding some 5 per cent per annum to the gross domestic product.

The impact on the changing structure of the Hong Kong economy has been considerable, with an increase in the importance of the service sector, although manufacturing remains by far the biggest employer (45 per cent of the work force in 1976). There has been a rapid increase in per capita income to $US2590 in 1977, with Hong Kong now ranking third in Asia after Japan and Singapore.

Hong Kong thus represents an example of rapid modernization within a densely packed urban environment. This has been accomplished within a political framework that has only limited political participation by most citizens. Because of the economic success of Hong Kong and the limited demand of non-urban areas, the Hong Kong government has been able to make considerable investment in the improvement of the urban environment and the provision of urban services.

KOREA

The Republic of Korea is another example of rapid economic growth and modernization. Between 1963 and 1976 Korea grew to be one of the most successful of the world's developing economies. The reasons for this growth are numerous, including existing historical structures which facilitated economic development.

Just as significant was the development approach adopted by the government of Chung Hee Park, which assumed control after a military coup in 1961 and continued in power until his assassination in 1979. Park introduced a series of economic reforms in the early 1960s with twofold objectives: first, to direct Korea to increased export-led industrialization, and second, to increase both private and public savings. These aims were accomplished by creating an institutional and trade environment which allowed unrestricted access to imported inputs as well as exemptions from tariffs and indirect taxes, and by adjusting the official exchange rate upwards to reflect more realistically the value of the national currency. Bank interest rates were also increased in order to increase private savings. Finally, foreign aid and international investment were encouraged to help fund the infrastructure and industries being constructed, particularly in the period of the late 1950s, when it made up 80 per cent of the investment.

The results of these policy initiatives were dramatic. Part of the growth was certainly due to the government export promotion programme, in which the government relied on market incentive policies to ensure that firms made an adequate profit, but the private sector has still had to take risks in pursuing these incentives.

The results of this liaison between government and the private sector have greatly affected the patterns of urbanization and class formation in Korea. Between 1960 and 1970 industrial employment more than tripled and accounted for 44 per cent of the total job increase between 1960 and 1970 and a large proportion into the 1980s. The contribution of manufacturing to the GNP rose from 13 per cent to 22 per cent in 1971 and 28 per cent by 1980. Urbanization levels also increased from 28 per cent in 1960 to 51 per cent in 1975, which meant that Korea was as urbanized as the average developed countries shortly after the

Second World War. Much of this increase was concentrated in the rapidly growing metropolis of Greater Seoul, which rose from 2.4 million in 1960 to 5.5 million in 1970, absorbing 42 per cent of the total urban increase. By 1980 it had risen to 8.4 million. In 1983 the population was said to have risen to more than 11 million. Despite this, in the period since 1970 the fastest growing areas have been in the southeast, particularly around Ulsan and Pusan where much heavy industry has been established. This trend seems likely to continue.

As Table 5.2 shows, the rapid urban increase has been sufficient to make Korea 50 per cent urbanized in a very short space of time, causing many problems of urban overcrowding and urban service provision, particularly for the lower-income communities.

MALAYSIA

In Malaysia there has also been rapid economic growth, reflected in the urbanization level shown in Table 5.2. Malaysia has also experienced an industrial revolution of sorts, but as yet it has not been of the size of Korea's. The growth of manufacturing has been largely concentrated in the major urban areas of Kuala Lumpur, Penang and Johore.

More precise information provided by Ariffin (1981) points out that manufacturing has been one of the fastest growing sectors in the peninsular Malaysian economy. Utilizing data from the 1976 labour survey, Ariffin shows that between 1957 and 1976 employment in the manufacturing sector grew dramatically by more than 320 per cent. Most significant has been the increase in female labour which has seen the proportion in the manufacturing labour force grow from 17 per cent in 1957 to 41 per cent in 1976.

In the context of Malaysian political economy this is a revolution of major consequence, for it involves the introduction of a growing number of previously rural dwellers into urban factories. The growing number of Malay female workers is also of major importance. The most significant increase in the racial mix of this labour force has been the growth of Malay female workers who have increased their numbers in manufacturing from 9000 in 1957 to some 110,000 in 1976. This means that there are almost as many Malay female workers as the Chinese, who are much more highly urbanized. The impact of the industrial revolution has also led to rapid growth in the major industrial centres of Penang, Greater Kuala Lumpur and Johore Bahru. Indeed Greater Kuala Lumpur grew from less than 350,000 in 1957 to more than 1 million by 1980.

PHILIPPINES

In the case of the Philippines, economic growth has been much slower, as indicated by the lower per capita incomes, despite considerable effort on the part of the government to industrialize the economy and increase rural productivity. At least one important reason for this is the continuing high rate of population increase (2.71 per cent) during the 1970s, leading to an increase in the total population of 11,414,000 people, which has seriously affected programmes of economic development.

This rapid rate of population increase has also been associated with quite rapid urbanization, lifting the urban population from 32.9 per cent in 1970 to 37.5 per cent by 1980. A large part of this population increase was due to migration (55 per cent in metropolitan Manila). This urban growth is not spread evenly throughout all urban centres for it is heavily concentrated on the metropolitan Manila area with 32.9 per cent of the national urban population in 1980. The dominance of metropolitan Manila is illustrated by the fact that almost two-thirds of the country's manufacturing establishments and a major proportion of the educational and governmental institutions are located in this area. The population of metropolitan Manila is almost four times bigger than the combined total of the next four largest cities, namely Cebu, Davao, Zamboanga and Bacelod.

The consequences of this rapid urbanization, particularly in metropolitan Manila and the surrounding regions of southern Luzon, have placed great strains upon the provision of employment and urban services, resulting in inadequate transport, continuing widespread poverty and the growth of squatter communities. At the institutional level the Philippines has been very innovative in the face of these problems, creating a 'Metro Manila Commission' in 1975 which integrated the various cities and municipalities of the Manila region into a unified metropolitan government. A 'Ministry of Human Settlements' was created in 1978 to co-ordinate activities in three main areas: (a) environmental management, (b) land-use and town planning and (c) shelter system development.

Despite these excellent institutional advances the Philippines is still faced with continuing economic problems (exacerbated in the last 2 years) which are particularly grave with respect to labour absorption in the urban areas. As a consequence, the informal sector is highly developed in the majority of Philippine cities. The Philippines therefore lies somewhere between Indonesia and Malaysia, Korea and Hong Kong in its socio-economic setting in its urban trajectory.

INDONESIA

Indonesia, the fifth country, presents the most intractable of problems. It possesses a very large population, 150 million (1980), highly unevenly distributed, with over 60 per cent on the island of Java, which is increasing in population at over 2 per cent per annum. Much of this population is still engaged in agriculture (60 per cent in 1977), with a high proportion in various forms of rice agriculture on the island of Java. As Table 5.2 shows, Indonesia is by far the poorest of the five countries and there are considerable differences between urban and rural income, and between high and low income groups. Despite the overall prevalence of poverty, Indonesia has experienced a number of major socio-economic changes over the last 20 years (not least has been the sizeable contribution of oil production to national revenue) which has enabled the government to hold-off the growing demographic pressures. In addition, Indonesia has apparently experienced an accelerated rate of urbanization between 1960 and 1980 (3.6 per cent 1961–70 to 4 per cent 1971–81, average annual growth rate), which is almost double that of the rate of increase of the total population. This process has increased the level of Indonesian urbanization to approximately 25 per cent. Of even greater significance is the fact that this urban growth has been largely concentrated on the larger cities above 500,000 in size, such as Greater Jakarta, Surabaya, Semerang, Bandung, Medan, Palembang, and Ujung Pandang. These are the urban areas where manufacturing, employment and production have increased substantially in the last decade and are important poles attracting migration. Thus World Bank data suggest that 48 per cent of the total net migration to urban areas between 1970 and 1981 went to Jakarta.

A second development of no less importance has been the introduction of new high-yielding varieties of rice and new technologies of agricultural production which have led to greatly increased yields in sawah agriculture. Hopes that these increases in agricultural productivity might hold the population in rural areas have not been fulfilled. Thus agriculture, which accounted for 76 per cent of rural employment in 1971, absorbed only 8.5 per cent of the total increase in rural employment. The rest was apparently absorbed into low-productivity service and industrial activities. The increase in rural non-farm labour has also been characterized by increases in circular migration, facilitated by greatly improved transportation at relatively cheap prices.

These developments certainly suggest that if existing trends continue, then Indonesia will experience increasing urbanization over the next 20 years which certainly in Java may lead to the creation of some extremely large urban areas where a sizeable proportion of the population will be low-income dwellers in need of urban services.

While it is clear that these divergent urban trajectories have been brought about by a complex mixture of resource endowments, demographic characteristics, historical patterns of incorporation into the international economy, types of technological development and particular development ideologies adopted by countries, it is remarkable how in a short period of time the Asian market economies have begun to show patterns of convergence and divergence similar to those of Latin America.

In both Asia and Latin America the processes of convergence (illustrated by the growth of industrialization and urbanization in countries such as Argentina and Korea) and divergence (in the stagnating rural economies of Bolivia or India) are the result of the common experience of incorporation into the world system. In both regions regional cities, such as Singapore and Rio de Janeiro-São Paulo, are emerging as beachheads in this process and a new class of transnational decision-makers is planning and implementing the process.

Capital circuits and networks

Despite the realization that urban patterns of individual countries are at least partially shaped by the flows of international capital, there has been surprisingly little focus on how this process operates in the developing world.[4] David Harvey's[5] interpretation of this process is of considerable help in providing a theoretical framework for this type of analysis. As he points out, capital can be seen to be operating in three circuits. First there is the circuit of *primary capital* which involves the appropriation of surplus value from labour. Second, there is the circuit of *secondary capital* which involves the flow of capital into 'fixed assets', where it is important to distinguish between fixed capital engaged in the production process and that which functions as a physical framework for production (the 'built environment' for production). Within the secondary circuit, capital also flows into a consumption fund which is 'formed out of commodities which function as aids rather than direct inputs to consumption' (Harvey 1981, 96). Some items are directly enclosed within the consumption process (consumer durables such as cookers, washing machines, etc.) while others act as a physical framework for consumption (houses, pavements, etc.) – the built environment of consumption. Some items in the built environment, such as the transport network, function directly for both production and consumption. Third, there is a *tertiary circuit of capital* which consists of investment in science and technology and social expenditures designed to improve the processes of reproduction of labour power (e.g. education).

Harvey's main purpose in this analysis is to show how crises within these various circuits of capital can be resolved by switching of capital between the circuits, in particular in cases of overaccumulation in the primary circuit. This switch of capital can only be created if 'the various manifestations of overaccumulation can be transformed into money capital which can move freely and unhindered into these forms of investment' (Harvey 1981, 97). This needs a money supply and credit system creating 'fictional capital' in advance of production and consumption in both the consumption fund (housing mortgages, etc.) and fixed capital. Thus financial and state institutions act as 'a kind of collective nerve centre governing and mediating the relations between the primary and secondary circuits of capital' (Harvey 1981, 97).

This theoretical exploration is set within the empirical context of the analysis of long waves of capital accumulation and contraction in the world economy (see Kondratieff 1935; Mandel 1980; Frank 1980, 1981; Bergeson 1982). In this particular approach the long boom between 1945 and 1973 is now said to be entering a phase of contraction, involving, in Bergeson's view, a period of decline of the American hegemony and a form of merger between state and private corporation which will undermine the power of multinationals (Bergeson 1982). Perhaps of more interest is how this period of boom, which might be more accurately extended to the end of the 1970s (1973 being the important date of the start of the downturn) has precipitated a 'switching crisis' and has involved the spatial movement of capital as one response to the problem within what may be labelled networks of capital.

Broadly speaking, these flows of capital (which will be equated with financial resources) may be divided sectorially and geographically. Sectorially, three sectors characterize financial flows (capital): (1) Official Development Assistance, such as bilateral grants; (2) private flows, consisting of direct investment, bilateral portfolio investments, multilateral portfolio investments and export credits; and (3) grants by private voluntary agencies. During the period between 1960 and 1976 there has been a marked shift in financial flows from developed to developing countries, in that sector 1 has declined from 60 per cent to 42 per cent of the total, while private flows have increased from 38 per cent to 54 per cent. In this paper we are concerned primarily with private net direct investment flows and in the period 1969–76 the following picture emerges for Asia (see Table 5.3).

What is significant here is the very high proportion of investment in Indonesia (mostly in resources) and the quite significant investment in Hong Kong and Singapore (16.9 per cent) which contain only a small proportion of the region's population. It was this investment in these two city states which in part accelerated their roles both as manufacturing centres and important financial centres for the region. For the other countries much of this investment went into resources (petroleum, minerals) and manufacturing, and 90 per cent of the net financial flows in the 1960–76 period was contributed by five countries (USA 50.7 per cent, UK 9.5 per cent, Germany 8.8 per cent, France 7.7 per cent and Japan 7.5 per cent). From the point of view of sequential direction these five main contributors were quite different. Eighty per cent of the USA's net financial investment went to Latin America between 1966 and 1976 while over 47 per cent of Japan's investment went to Asia.

Geographically then we may recognize that at the level of direct net foreign investment there is an international network of capital flowing into the Asian

Table 5.3 Regional distribution of private direct investment flows in Asia, 1969–76

	$US millions	%
India	343	5.62
Pakistan	15	.25
Hong Kong	633	10.36
Indonesia	2869	46.97
Republic of Korea	594	9.73
Malaysia	526	8.61
Philippines	567	9.28
Singapore	402	6.58
Thailand	159	2.60
Total	6108	100.00

Source: Billerbeck and Yasugi (1979).

market economies dominated by the USA, Japan, Germany, France and the United Kingdom. The role of the international banking system is crucial in this flow of capital. Its importance in the Asian region may be gauged by the dramatic increase in the number of international banks active in the region. For example, in Singapore the number of international trading banks and investment banks increased from 60 in 1970 to more than 100 by 1983 (see Cushman 1976, Burn 1978, Sampson 1981). Within the region there are also networks of capital that may be labelled regional networks, in which the overseas Chinese (in particular) are operating, mainly out of Hong Kong and Singapore. As Wu and Wu point out 'there is ample reason to believe that over the years sizeable amounts of capital owned by ethnic Chinese in Asia have been deposited' (Wu and Wu 1980, 94) in these city states awaiting reinvestment. For instance 'Hong Kong financial authorities have estimated the annual flow of Chinese funds from Southeast Asia since 1964 at $HK1 billion ($US80 million a year)' (Wu and Wu 1980, 95) and Singapore has played a similar role through its Asian dollar market (see Table 5.4). This network of capital is facilitated by a financial network in which overseas Chinese banks play a highly important role. Taiwan also plays an important role in the flow of capital as is shown in Table 5.4.

Third, there is the capital network at a national level where national capital is invested within the nation. Finally there is the sub-regional capital network where provincial capital is invested within a province.

These geographically defined networks are, of course, artificial, for financial flows recognize no boundaries except those imposed by political intervention.

Table 5.4 Foreign investments in selected Southeast Asian countries by investors in Hong Kong, Singapore and Taiwan ($US millions)

	Total from all countries	*Hong Kong*	*Singapore*	*Taiwan*
Indonesia[1] (1968–75)	2828.3	267.4	100.5	16.0
Philippines[2] (1968–75)	500.7			85.1
Thailand[3] (1960–75)	197.1	11.2	2.3	22.6
Malaysia[4] (1966–75)	535.5	26.0	144.5	2.8

Notes: [1] Approved investment projects; [2] approved investment projects; [3] total required capital; [4] total approved investment.

Source: Wu and Wu (1980), p. 102.

In simple terms what this indicates is a substantial growth in the fluidity and availability of capital (whether through direct investment or indirect investment) in the market economies of Asia, the flow of which has been greatly facilitated by the growth and development of a wide variety of financial institutions. Part of the 'switching crisis' of the developed countries has been resolved by a geographical shift in investment to the developing countries. How has this facilitated and impacted on the urbanization process and labour force formation process in Asia?

A comparison of the diverse strategies adopted by the Japanese in investment in Malaysia and Indonesia indicates how the flows of capital affect internal patterns. In the early 1970s Japanese business and government authorities made decisions designed to slowly restructure the Japanese economy over the next two decades. This called for exporting low value-added manufacturing and high pollution manufacturing processes 'off-shore' while continuing to develop high technology industries in Japan (see Table 5.5). In Malaysia this led to a very rapid growth of Japanese investment in electronics and labour-intensive firms. By 1977 there were 489 firms with a culminative capital of $US425 million located mostly in the developed western states. During the 1970s there has also been investment in natural gas and other developments by firms. This reaction to a 'switching crisis' by Japan has thus been the basis of a remarkable growth in manufacturing in Malaysia, shifting the proportion of people engaged in industrial activity from 12 per cent in 1960 to 20 per cent in 1977 (McGee 1982, Golley 1983). Japanese and other foreign and national investment in industry has also influenced the urban system quite markedly.

In the case of Indonesia a different pattern of Japanese investment has been exhibited, with over two-thirds of its investment up to 1977 in the mining sector, much of it in some form of joint venture with Indonesian state or private

Table 5.5 Number of Japanese firms operating outside Japan by selected sector in which principally operating and year of establishment

		1970–4		*1975–9*		*1980–1*	
		No.	*%*	*No.*	*%*	*No.*	*%*
World	Manufacturing	935	37.6	705	37.1	364	38.7
	Commerce	789	31.8	690	36.4	329	35.0
Asia	Manufacturing	640	60.3	410	57.9	177	57.8
	Commerce	174	16.4	156	22.0	68	22.2
US	Manufacturing	60	12.8	89	22.3	78	31.2
	Commerce	262	55.9	217	54.4	104	41.6
Europe	Manufacturing	46	13.9	60	19.1	42	22.7
	Commerce	197	59.5	191	60.8	107	57.8

Note: Work is proceeding on this data base which also includes information on start-up capital. When this is processed it will greatly increase the importance of mining and manufacturing. This includes all companies in which Japanese companies have some form of participation.

Source: Oriental Economist, 9 December 1982. Based on data collected by Tamiko Kurihara.

companies, to the extent that Indonesia is now Japan's most important trading partner. A high proportion of exports are made up of oil and minerals, but in the period since 1975 there has been more investment in processing industries such as the Asahan Aluminum Smelter (see Rocamora 1983; Forbes 1982). One consequence of this pattern of investment has been that Indonesia's urbanization has been slower and it has experienced a slower growth of the industrial labour force.

Table 5.5 provides a broader overview of this process operating at a world level. This table shows how the Japanese strategy of off-shore investment has concentrated on manufacturing in the Asian region and the development of commercial activities in the high-income markets of North America and Europe.

Another response to the 'switching crisis' is to invest capital in the built environment of consumption, such as offices, hotels, transportation and residential building, etc. This can be greatly facilitated by the ability of financial institutions to loan money and by a state legislation and monetary policy which encourages this activity. During the decade of the 1970s (particularly in the late 1970s) virtually every major Asian city experienced a very rapid growth in office and hotel buildings which transformed the central business districts of these cities into mini-Manhattans, much of this the result of generous loans from the banking

Table 5.6 Hong Kong: sectoral distribution of domestic bank loans[1] ($US millions), 1975–82

	1975	*1978*	*1982*
Manufacturing	11.9	12.6	9.5
Agriculture, fisheries	0.1	0.1	0.1
Transport, transport equipment	9.0	7.9	8.9
Electricity, gas and telephone	1.0	0.5	1.4
Building and construction	7.2	10.0	27.9
Wholesale and retail trade	37.2	29.4	5.8
Miscellaneous	33.5	39.4	46.3

Note: [1] Includes deposit-taking companies. Domestic banks raised funds overseas for domestic loans.

Sources: Lee and Jao (1982), p. 241; Hong Kong Government (1983), p. 55.

system and a combination of both national and regional capital (particularly that of the overseas Chinese). As one writer in *Far Eastern Economic Review* commented, 'Overseas Chinese . . . seem to have an innate faith in the ownership of bricks and mortar' (Kulkarni 1983, 71).

Table 5.6, which shows the sectoral distribution of domestic bank loans at the ends of 1975, 1978 and 1982 in Hong Kong, indicates the dimensions of these assertions, with domestic bank loans for building and construction rising from 10 per cent of total domestic bank loans in 1978 to 28 per cent by the end of 1982. This boom was shared in other Asian countries. For instance, in Indonesia bank credit extended for construction and manufacturing (available statistics do not separate the two categories) grew from 25 per cent of the total credit in 1975 to 51 per cent of total credit extended in 1980 (see World Bank 1982). Of course, it is not possible to give an accurate estimate of what proportion of this credit was utilized for the built environment of the inner cities but it can be asserted that a considerable proportion was used for this purpose. For instance, in Indonesia in 1980, Jakarta, with 3.8 per cent of the population, was the source of 48 per cent of Indonesia's extended bank credit.

Of course the glitter of this property boom has now largely gone, particularly in Hong Kong (where geopolitical factors aggravated a deteriorating economic situation produced by a world-wide recession). Even in Singapore, which has maintained a reasonable rate of GNP growth during the recession, Kulkarni reports that

> projects already under construction and due to commence soon indicate a massive oversupply by 1986–7 in all sectors – residential, office, commercial

> and industrial. Total transactions in the property sector in 1982, amounted to $S616 million ($US290 million), a reduction of 38.7 per cent by value from the 1981 high of $S1.59 billion. (Kulkarni 1983, 71)

This situation, as with many other Asian countries, had its origins in the late 1970s when very rapid rates of economic growth, together with the government's own modernization programmes, encouraged many developers to invest in the property market. Banks, pursuing Asian dollars and other regional business, as well as multinationals snapped up the space made available by property development. In this section we have analysed in some detail how various types of capital flows affect urban patterns of selected Asian countries. In the next section we deal with another sectoral aspect of these capital flows, namely the growth of investment in industry and how this has affected the urbanization process in Asia.

Urbanization and industrialization: the main components of this relationship in the Asian market economies

One of the most persistent generalizations concerning the urban transformation in the developed countries is that it rested upon a series of technological and productive changes which led to a large growth in industrial and agricultural production. These changes led to a change in the occupational composition of the population from agriculture to a dominance of non-agricultural pursuits and an increasing concentration of people in urban places. While there was great variation in the temporal sequence of this process in Europe there is no doubt that this process was the basis upon which the urban transformation was accomplished.

It has been claimed that this relationship between industrialization and urbanization is very different in the Asian market economies. The historical incorporation of these countries into the world economic system has created a sharp division of economic activity, in which the Third World countries have become largely suppliers of raw or semi-processed materials to the developed countries and importers of manufactured goods from them. There is little development of indigenous industry; this has meant that the urbanization process is very different, with a growth of large urban places which act as focuses in the funnelling of these commodities to, and from, Third World countries. The occupational structure of these cities is highly concentrated in the tertiary sector and overall levels of urbanization remain low. Rapid population increase resulting in rural-urban migration and high rates of natural increase in the cities have further exacerbated this situation, leading to considerable pessimism as to the likelihood of urban transformation in many Third World countries.

While this contrast may have been true of the market economies of Asia in the 1950s, there is increasing evidence that this situation is undergoing considerable change. In the period between 1960 and 1980 industrialization has been growing rapidly in many of these countries. Table 5.7 shows that virtually every Asian market economy has experienced a sizeable increase in the contribution of industry to the Gross Domestic Product. Care should be taken in interpreting these figures, however, for they include many different types of industrial production which have different locational and employment effects.

This growth is also reflected in Tables 5.8 and 5.9, which show changes in the percentage share of merchandise exports and imports between 1960 and 1981. Thus, in the rapidly industrializing countries of Hong Kong, Singapore and Korea, exports of manufactured goods became a major proportion of total exports. In other countries the increase was less and the percentage share of primary commodities fell substantially, indicating the structural changes occurring in the Asian market economies.

Table 5.7 Selected Asian market economies: structural contrasts in the distribution of the GDP, grouped by urbanization trajectories, 1960–80

	Agriculture		*Industry*		*Manufacturing*		*Services*	
	1960	*1980*	*1960*	*1980*	*1960*	*1980*	*1960*	*1980*
Type I								
Hong Kong	4	2	39	31	26	26	57	67
Singapore	4	1	18	37	12	26	78	62
Taiwan	28	9	29	46	22	42	43	49
Korea Rep.	37	16	20	39	14	28	43	45
Type II								
Malaysia	36	23	18	30	9	18	46	47
Philippines	26	22	28	36	20	24	46	42
Type III								
Thailand	40	22	19	28	13	19	41	50
Sri Lanka	32	27	20	27	15	15	48	46
Type IV								
Bangladesh	57	47	7	14	5	7	36	39
India	50	33	20	26	14	16	30	41
Pakistan	46	31	16	25	12	17	38	44
Indonesia	54	26	14	39	8	13	32	35
Burma	33	48	12	13	8	9	55	39

Source: see Table 5.1.

Table 5.8 Selected Asian market economies: percentage share of merchandise exports, 1960–81

	Fuels, minerals, metals		*Other primary commodities*		*Textiles, clothing*		*Machines, other transport equipment*		*Other manufacturing*	
	1960 %	*1981* %	*1960* %	*1981* %	*1960* %	*1981* %	*1960* %	*1981* %	*1960* %	*1981* %
Hong Kong	5	1	15	2	45	42	4	18	31	37
Singapore	1	29	73	15	5	4	7	26	14	26
Korea Republic	30	2	56	8	8	30	—	22	6	38
Malaysia	20	36	74	44	—	3	—	12	6	5
Philippines	10	16	86	39	1	7	0	3	3	35
Thailand	7	8	91	65	0	10	0	5	2	12
Sri Lanka	—	14	99	65	0	16	0	—	0	5
Bangladesh	—	0	—	32	—	56	—	1	0	11
India	10	8	45	33	35	23	1	8	9	28
Pakistan	0	7	73	40	23	41	1	1	3	11
Indonesia	33	83	67	13	0	1	—	1	—	2

Source: see Table 5.1.

These changes were associated with a shift in the percentage share of merchandise imports, emphasizing the growth of 'transformation' imported goods needed to provide the technological inputs for the growing industrialization and economic growth. Only India and Pakistan did not share in this change, owing to their own ability to produce 'transformation' goods. Most Asian market economies have pursued industrialization strategies by first encouraging import substitution, particularly in the production of consumer durables.[6] There was also an important distinction in the product mix in this process of import substitution. In a country like India considerable effort has been made to build up the iron and steel industry as a basis for the development of a significant capital goods industry; indeed India has become one of the major Third World producers of these goods (see Mitra 1979). South Korea entered by another path, beginning to develop its capital goods industry after its substantial development of lower order exports (see Koo 1980).

On the other hand, most other Asian market economies followed policies of import substitution of consumer durables which often involved assembly of imported components. This form of industry is highly concentrated in the large cities, close to the major market composed of a growing middle and upper

Table 5.9 Selected Asian market economies: percentage share of merchandise imports, 1960–81

	Fuels		*Other primary commodities*		*Food*		*Machines, other transport equipment*		*Other manufacturing*	
	1960 %	*1981* %	*1960* %	*1981* %	*1960* %	*1981* %	*1960* %	*1981* %	*1960* %	*1981* %
Hong Kong	3	8	16	5	27	12	10	23	44	52
Singapore	15	34	38	5	21	7	7	28	19	26
Korea Republic	7	30	25	15	10	12	22	21	52	37
Malaysia	16	17	13	5	29	13	14	37	28	28
Philippines	10	30	5	4	15	8	36	23	34	35
Thailand	11	30	11	8	10	4	25	26	43	32
Sri Lanka	7	25	5	3	39	19	15	23	34	30
Bangladesh	—	8	—	11	—	20	—	21	—	40
India	6	45	28	8	21	9	30	13	15	25
Pakistan	10	28	2	8	22	14	27	23	39	27
Indonesia	5	13	10	6	23	11	17	36	45	34

Source: see Table 5.1.

income class. Malaysia and the Philippines were very active in encouraging this type of development in the late 1950s and 1960s. It was an important factor in the growth of the major metropolitan areas in many Asian countries, although the high capital-labour ratio did not always lead to rapid labour absorption, particularly given the very rapid growth of the labour force. Finally, while both forward and backward linkages have developed for different products subject to import substitution policies (for instance, the Malaysian motor car assembly industry has an increasing local content), this has not been sufficient to achieve an industrial revolution in most Asian countries.

During the 1960s and 1970s there has also been a considerable growth of export-oriented industries in many Asian countries; this has important implications for the urban structure. There is little doubt that the value of manufactured exports from the less developed countries has increased substantially. In the period between 1965 and 1975 these exports increased from $US4.6 billion to $US33.2 billion at the rate of 16.3 per cent, compared to 10.8 per cent in the developed countries.[7] This increase was geographically most unevenly distributed, with over 50 per cent of it occurring in the major Asian export platform economies of Korea, Taiwan, Hong Kong and Singapore. Indeed for the period between 1970 and 1976 UNCTAD researchers have pointed out that this concentration was much higher for certain products. Hong Kong and Korea alone accounted for almost 70 per cent of the increase in that period of exports of clothing by the less developed countries to developed countries. For miscellaneous light industrial products (electronics dominate) the figure was almost as high.[8] While the product mix has been dominated by the textile, clothing and electronic products, there has also been an increase in machinery, transport equipment and chemicals. Finally, while these increases have been considerable, they only increased the share of the less developed countries' manufactured exports in the world total from 4.0 per cent to 7.4 per cent (Keesing 1979, 24).

In order to understand these developments, it is necessary to trace some of the major changes in the international economy which have facilitated them. These changes are quite diverse, and to a large extent account for the complex mixture of ownership and production processes that characterize the manufacturing export activities of the Asian market economies.

First, at the technological level, developments in the transportation of goods have greatly facilitated the growth of export industries in the market economies of Asia. Most important in this respect is the 'container revolution' and bulk air cargo carrying. Investments in the transportation infrastructure of Asian market economies, notably in port handling equipment and improved airport facilities, have enabled firms to take full advantage of these transportation developments.

Secondly, the type of commodity being exported has an important influence on industrialization in the market economies of Asia. For some products, particularly those associated with the electronics industry, it is possible to separate production processes rather than produce one integrated unit at a single location. It has been possible to put out the labour-intensive assembly parts of the production process into the Export Processing Zones of the market economies of Asia. This has been encouraged when production is under the control of large transnational corporations which control the whole process from the purchase of raw materials to the marketing of the product (see Helleiner 1975; Nayyer 1978). Finished products which rely on labour-intensive inputs (such as textiles, toys and shoes) are more likely to be produced by sub-contracting to domestic firms by large marketing firms in the developed countries (see Watanabe 1972; Sharpston 1973).

A third factor has been the tariff policies of the developed countries. The adoption of US Tariff Items 870.00 and 806.30 has been responsible for a considerable growth of export industry in the market economies of Asia (see Finger 1975). The special relationship of Hong Kong with Britain is another example. It should not be imagined, however, that the developed countries have lowered tariff barriers completely, and in the current recession there is much pressure to re-establish them. This will encourage Asian exporters to seek markets in other Third World countries.

Finally, there is the very important factor of relative wage costs. Ratios of US earnings to those of Asian market economies are often ten times for the same job. In addition, the low degree of unionization, often encouraged by the state, means that wage rates are rising only slowly in the Asian countries and the labour force is much less active in demanding improvements in its working conditions (see Helleiner 1973a, 1973b).

The growth of industrialization was not simply induced by the government policy of the various states. It was also the result of a considerable increase of investment by foreign countries, notably Japan. Table 5.10 shows the increase in Japanese investment by major manufacturing type in the period between 1970 and 1976. It is clear from this Table that a large proportion of Japan's direct investment went into early growth industries such as textiles and electronics which were most clearly associated with the growth of export-free zones. It should be noted, however, that this investment was regionally concentrated, with almost 50 per cent in the four most industrialized and urbanized countries of Hong Kong, Singapore, Taiwan and South Korea.

Table 5.10 Japan's direct foreign investments ($US millions)

	1970	*1976*
Japan's total investment	3,557	19,405
Total directed to Asia	780	5,464
All manufactures in Asia		
Textiles	108	718
Electric machines and appliances	45	294
Other manufacturing	49	291
Steel and non-ferrous metals	6	270
Chemicals	23	204
Transport equipment	14	136
Timber and pulp	13	113
Food processing	33	99
Heavy machines	15	97

Source: Kojima (1978).

Conclusion

This brief description of the urbanization process in Asia has emphasized four main points. First, the urbanization process in Asia is very diversified between countries. Second, there is a significant growth of capital flows within the region which is very significant in shaping the urban patterns of the region. Third, the growth of industrialization has an important impact on urban employment structures in several countries in the region. Finally, there are indications of the emergence of an international urban hierarchy in the region in which Tokyo, Hong Kong and Singapore play significant functions as global cities.

There are two views about the last development. The first view, favoured by researchers such as John Friedmann, argues that these processes of expansion of international tertiary activities are tending to concentrate in large global cities such as Tokyo, Singapore, Los Angeles, New York, etc. at the expense of cities which are primarily serving national service functions. In such a scenario an international urban network of cities would emerge, primarily generating income through the international services they perform. Central to Friedmann's arguments is his view that these cities would be the base of world multinationals, who would be controlling more and more of the world economy. The policy implications of the viewpoint are, of course, of extreme importance to the governments of the Asian region, for they suggest that they will have to devise programmes designed to encourage this international activity if these cities are to remain competitive in this international market. Clearly, within the region the Singapore government has already recognized these developments and has set about creating an institutional, legal and built environment that is conducive to encouraging international companies to invest in their country.

A second viewpoint lays greater stress upon the ability of national states in the region to mediate these processes in such a way that international companies may be involved in joint ventures with national companies (see Helleiner 1979), thus retaining a high service component of their activity within national units. This view is further reinforced by the revolution in telecommunication and computer technology, which many observers think will allow greater dispersal of decision making and office functions within both international and national urban hierarchies. Clearly, both viewpoints are speculative in that there is little empirical research to support either position.

For the majority of the Asian population who reside in the rural areas of Asia the development of an international urban hierarchy will have limited effect. Their future rests upon a successful rural revolution leading to increased agricultural productivity and prosperity.

6 Ecuador: accumulation and the producers

Introduction

The purpose of the following four chapters is to take up and illustrate some of the themes outlined in the theoretical chapters by reference to case studies in Ecuador, Hong Kong and Malaysia. The area chosen for the first study lies in the south of Ecuador and comprises three provinces: Azuay, Cañar and Morona Santiago (see Figure 6.1). We will not be dealing directly with a major metropolitan area, or even a well-known city. Nevertheless, the processes of capital accumulation and the principal actors involved are well illustrated in Cuenca, the main city of the region, with its rural and small market town hinterland, and in the region's relationships with national metropolises and corporate capital.

The present chapter is concerned with the producers of surplus – peasant farmers and craftspeople in the towns and the countryside, and urban workers in the modern sector of Cuenca, as well as the wide range of people in petty trading and services. While the evidence of inequality soon becomes apparent to the observer, the mechanisms which create it, and the efforts made by popular groups to confront and deal with it, may be less obvious. The purpose of this chapter is to open them more clearly to view.

The national context

Before looking at the three southern provinces in detail, we need to situate them in the wider context of Ecuadorian development. As we know from Chapter 4,

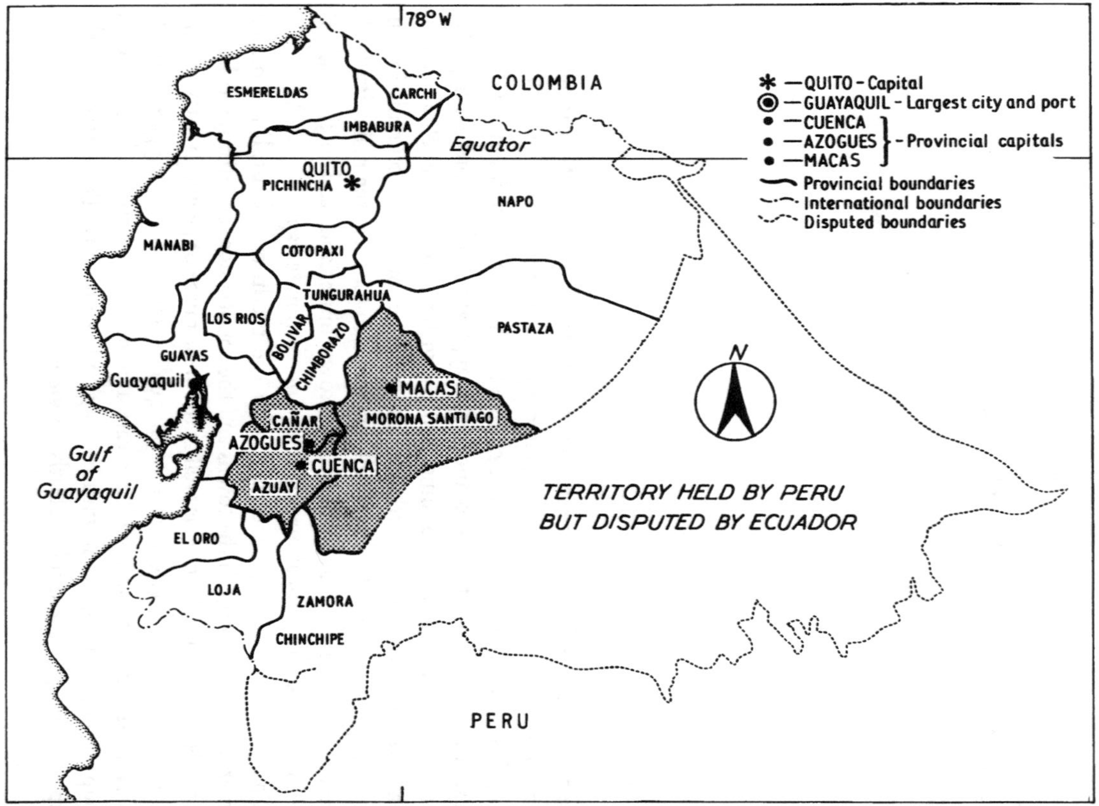

Figure 6.1 Ecuador: the southern region of Azuay, Cañar and Morona Santiago

Ecuador is one of the less industrially modernized societies of the continent. However, dramatic economic changes have followed from the exploitation of the oil reserves in the eastern region of the country since the early 1970s. Ecuador's population of 8 million is still predominantly rural (55 per cent in 1980 compared with 64 per cent in 1960). Population growth rates have remained high in the past 20 years, even rising in the period 1970–80 to 3.1 per cent a year. The rural contribution to this has been a yearly increase of 2.3 per cent, compared with 4.6 per cent in the urban areas. However, the numerical increase in rural population from 1960 to 1980 was not much less than the urban: 1.7 million compared with 2.2 million. The land, then, still provides direct support for a large and growing number of Ecuador's people.

The urban population increased by 140 per cent in the period 1960–80, from 1.6 million to 3.7 million, and, in the decade 1970–80, had one of the highest annual rates of increase in the sample, ahead of any other South American society. The degree of concentration in the largest city remained low at 16.8 per cent in 1970. But with rapid economic change in the mid-1970s, the percentages in the two largest cities, Guayaquil and the capital, Quito, increased. With a combined population of over 1.5 million, the result of rapid natural increase plus migration, they account for nearly 40 per cent of the urban population. By comparison, Cuenca, the third-largest city, has less than one-seventh the population of Quito and one-tenth that of Guayaquil.

The picture of a country still predominantly rural but undergoing change is reinforced by other indicators. Just over half the labour force in 1979 was engaged in agriculture, one-fifth in industry (much of this in domestic and 'lower circuit' activity). The distribution of domestic product shows that only 15 per cent was contributed by agriculture in 1979 (compared with 33 per cent in 1960), 37 per cent came from industry (of which 19 per cent comes from the modern manufacturing sector), and nearly half from the service sector.

The dramatic sectoral changes that have occurred in the past decade revolve around the exploitation of the oil in the Oriente province of Napo. In 1978, fuel exports accounted for over 40 per cent of the total (compared with none in 1960), while primary commodities had dropped from 99 per cent in 1960 to 57 per cent in 1978. On the import side, nearly 90 per cent of all products were machinery and transport equipment (50 per cent) and other manufactures (38 per cent) in 1978.

Ecuador is, historically, the classical agro-export economy. Over the past century it has produced a series of cash crops for international markets – in the nineteenth century for western Europe, then, later, for the United States market. In succession, various export crops came to dominate the economy; with world market changes, each declined and, after a period of difficulty and instability, ceded place to yet another export product (Velasco 1973a, 1973b).

Cacao was the source of wealth for the landowners of the coastal provinces and for the merchant-bankers of Guayaquil until the early 1920s. It was replaced in the depression years by coffee, which, in turn, gave way to rice in the war years (Ecuador now imports rice). A few lean years were followed by the banana bonanza decade of the 1950s. The United States' fruit corporations, their plantations struck by Panama disease, moved their centre of operations from marginal Central American and Caribbean exporters to Ecuador, and converted the coastal area of the country into large banana plantations. That particular boom lasted until the 1960s when the corporations switched back to their traditional Central American plantations with the new resistant Cavendish strain. Ecuador then reverted to the position of reserve supplier to the United States market (Larrea-Maldonado 1982, 28–34).

The 1960s were characterized by instability and uncertainty as the dominant classes, especially those in banking and commerce, endeavoured to adjust to the change. The early 1970s, however, brought the commercial exploitation of the oil reserves discovered in the Northern Oriente region. Then Ecuador entered on yet another of its export-fuelled booms; economic growth surged as the oil was exported at the rapidly increasing OPEC prices.

This long-term boom/bust pattern has had deep social and political consequences, for the commercial export groups have usually commanded political and administrative power, either directly or indirectly, through military intermediaries. National policies have therefore tended to stress the free market development of an economy open to international corporate trade and investment activity.

Regional inequalities have also been exacerbated by the expansion of the coastal cash crop economy. The principal beneficiary has been the city of Guayaquil, centre for the commercial and financial élites, and the branch offices of foreign companies. Only since the oil boom has Quito, the capital and increasingly a financial and industrial centre of operations for foreign capital, come to challenge Guayaquil's dominance. The consequent growth 'corridor' between the two cities is drawing into it investment and economic activity, together with a significant proportion of the national workforce.

The military junta, which took control in 1972, committed itself to a policy of modernization and structural reorganization financed by large petroleum exports, a progressive nationalist programme of social, sectoral and regional reconstruction which would ensure economic growth and social transformation to a modern progressive bourgeoisie, and greater political stability.

The effect of the supply side policies has been to transfer resources to national and foreign capital through credits, subsidies and tax exemptions. While total domestic product expanded dramatically up to 1976, especially, as a result of oil exports, the net effects of social and regional reconstruction programmes have been minimal. 'Despite the state's efforts, only an acceleration of the growth rate was achieved, while the historic heterogeneity of the socio-economic formation and its dependency were deepened' (Bocco 1982, 54).

Demographic and economic growth concentrated even more around the two urban poles of Quito and Guayaquil. The consequences of this are visually obvious in the new skylines, the multi-lane traffic arteries and in the new North American-style suburbia. They are also apparent in the shanty towns on the estuaries of the Guayas River around the outskirts of Guayaquil, which house the migrant populations descending from the Sierra provinces.

The two cities are the focal points of the national modernization strategy. The headquarters of national banks and their foreign counterparts are spread between them, as are the commercial and industrial establishments. If Guayaquil has for some decades enjoyed the advantages of being the major port for an export economy and the centre of a rich export crop hinterland, Quito has recently begun to grow more rapidly as a consequence of its role as national capital and its location near the oil-producing regions.

The multiplier effects from petroleum sector growth can also be seen in the emergence of a modern manufacturing sector in both cities. Whether such modernization represents a break with the dependent growth patterns of the past is another matter, for the reliance upon foreign capital, technology and management has deepened, and the external debt has heavily increased. (Bocco 1982, Table 4). Ecuador is a prime case of 'dependent industrialization', or, to use the phrase of Tavares and Teixeira, of 'internationalization' of the manufacturing sector.

The consequences of this type of growth are those discussed in Chapter 4: above all, a capital-intensity of production which limits employment opportunities in the sector, but provides manufactured goods for a local market increasingly influenced by the consumer preferences of North America and Europe.[1]

The dominant classes of Quito and Guayaquil stand at the centre-point of this derivative strategy, concentrating both modern production and consumption around their own interests at the expense of the socially and regionally less privileged, and sustaining this inequality through their political influence (Bocco 1982, Table 8). Some of the implications of this uneven pattern of national development for the southern region of Ecuador will become evident in the following sections.

Accumulation and producers in southern Ecuador

But not MONEY
    and because there was not money
        there was neither prostitution nor robbery

the doors of the houses were left open
neither was there Administrative Corruption nor embezzlement
–every two years
they accounted for their actions in Cuzco
because there was neither commerce nor currency
there was no
sale of Indians
never was an Indian sold
And there was chicha for all.
(Ernesto Cardenal 1972)

THE AUSTRAL PROVINCES

The area in the study comprises three southern provinces of Ecuador. Two, Azuay and Cañar, are predominantly provinces of the Sierra, spread over the rugged double spine of the cordillera which rises on either side of the narrow and sinuous Andean valley. Most of the population lives along this valley at altitudes of 2000 metres and above, with the poorer groups pressing further up the hillsides almost on to the *páramos* (high moorlands), and spilling over into smaller subsidiary valleys. There is fertile land in the central basins and some of the smaller valleys, but the density of population and the uneven property distribution dictate an intensive use of the hillslopes at altitudes often above 3000 metres.

The third province, Morona Santiago, is one of Ecuador's four provinces of the Oriente. It covers most of the foothills of the eastern cordillera and the headwaters of the Marañon River (which eventually runs into the Amazon), and is the region's resource frontier. Given the poverty of the Sierra provinces with their predominantly rural populations, it is not surprising that Morona Santiago is seen as an El Dorado waiting to be tapped by the combined efforts of private capital and national government planners.

The problems facing the region are defined as 'underdevelopment' (the lack of modern commercial and industrial structures and economic growth) by the local ruling class, government officials in Quito and international development planners. Yet it was not always so. Cuenca, with a population of 110,000, the urban focus for the three provinces, has a significant history. Before the Spanish arrived, Tomebamba was an important city in the empire of Tahuantinsuyo. It was from here that the Inca Huayna Capac directed the northward extension of his empire almost to the frontiers of present-day Colombia, and to the Ecuadorean coast. After the Spanish conquest the region was known both for its silver lodes which attracted the conquerors, and its *obrajes* or textile and leather works which supplied the even wealthier mining areas of Peru and Bolivia.

This internal imperial commerce brought wealth to the Spanish and their *criollo* successors until the mid-eighteenth century. But it also led to the exploitation of Indian labour in the mines and on the land. Indian communities entrusted by the Spanish Crown to *conquistadores* by means of the *encomienda* quickly degenerated into the classic *latifundio-minifundio* complex centred around the power of the creole landowner (Barraclough and Collarte 1973, Chapter 1).

Up to the last quarter of the nineteenth century, a closed, land-owning Sierra oligarchy governed both the colony and republic. Few social or political changes ensued before the 1890s, when the revolt of the merchant, financial and plantation owner export groups of the coast brought Guayaquil-controlled liberal regimes to power. Until recent years, this division between a conservative Sierra with its *latifundio-minifundio* structure, producing for the local market, and an export-oriented capitalist economy on the coast, has spatially characterized the unequal historical evolution of the country.

Azuay and Cañar have been part of this traditional society, but Morona Santiago's real incorporation into modern development dates back only a few years. The decline in mineral production and the *obrajes* left the Sierra provinces with minifundios, craft activities and large landholdings worked under the serf-like conditions of *huasipungo* by the bonded indigenous population. Primitive accumulation at the expense of the *huasipungueros* and unequal trading (price) relations between Sierra and coast have caused a pauperization of the peasantry. More recently, it has provoked a flow of migrants seeking employment in Guayaquil and the commercial agriculture of the coast.

The process is classic: conquest; expropriation of surplus; dispossession and forced migration of dislodged producers to seek work in the growth sectors of the export economy. Underdevelopment was not the natural state of the two Sierra provinces. It is a phenomenon that has gradually encroached upon the lives of the indigenous communities.

As exploitation deepened its impress over the centuries, it added a cultural dimension to economic impoverishment. Opposition has become covert, disguised; there has been a retreat into unresponsiveness, perceived by modernizers and development planners as peasant ignorance and conservatism. But such attitudes are evidence rather of a form of self-defence; as Gerrit Huizer has observed, 'rather than being a symptom of a natural or traditional incapacity and unwillingness to change, [they] can best be explained as the result of a long period of repression' (Huizer 1973, 2).

Equally, though, this centuries-long process of exploitation contains deep contradictions. As Huizer says:

> The overall presence of distrust and resentment among peasants under the traditonal *hacienda* system, in addition to the frustrating effects of modernization,

> seem to make the peasants ready for militant organization as soon as they can find sympathetic and able allies to guide and support them. (Huizer 1973, 5)

The forms of repression and exploitation which once isolated the peasantry, hemmed them in, and restricted their vision of the world, have given way to a socio-economic system which throws them on to the highways and byways of their society; it pushes them from their land into the cities and from Sierra to coast; it brings them into contact with other people, other cultural pressures and other lifestyles. These are experiences which frustrate and confuse, but which also lead peasants to question, and, in some cases, to reject. Examples of changing attitudes and their causes are presented later in the chapter.

SOCIAL AND ECONOMIC INDICATORS

The social and economic facts which document this situation are presented in the following pages. As seen in Table 6.1, the majority of the population is still classified as rural, although such a classification does not necessarily mean that all those who live in non-urban areas gain their livelihood there. Table 6.2 indicates that the three provinces are even more heavily rural than Ecuador as a whole, which is still a rural society. Moreover, if we compare Tables 6.1 and 6.2, it is apparent that, in Cañar and Azuay, there is a relatively heavy density of population to be supported in an economy that is dependent on peasant agriculture, craftwork and the multitude of semi-urban occupations outside the modern capitalist sector. This is particularly true of the interandine valleys. With less than 10 per cent of the total area of Cañar and Azuay, they contain over 90 per cent of the population (CREA (Centre for the Economic Reconversion of Azuay, Cañar and Morona Santiago, Cuenca), 1977, vol. 1, 21).

Table 6.1 Provincial areas' population, 1974

Province	*Area (km^2)*	*Population*		
		Census (estimate)		*CREA (estimate)*
		Total	*Rural*	*Total*
Azuay	8,300	384,278	262,550	365,657
Cañar	3,400	145,411	125,891	147,463
Morona Santiago	25,000	53,312	44,059	50,406
Total	36,700	583,001	432,500	563,526

Sources: CREA (1976) *CREA: Desarrollo Regional 1975; III Censo de Población: II Censo de Vivienda, 1974* (muestreo), Quito (1975).

Table 6.2 Urban and rural population, 1974

Province	*Urban*	*%*	*Rural*	*%*	*Total*
Azuay	121,728	31.7	262,550	68.3	384,278
Cañar	19,520	13.4	125,891	86.6	145,411
Morona Santiago	9,253	17.4	44,059	82.6	53,312
Ecuador (comparison)	706,677	41.3	3,845,369	58.7	4,552,046

Sources: Based on *III Censo de Población.*

The population grew rapidly between the censuses of 1962 and 1974 (Table 6.3). Although the differences in calculation vary because of a variety of data sources, it is clear that there was a high rate of natural population increase from 1962 to 1974, and only a relatively heavy out-migration from the region kept the increases below the national annual rate.

The urban population of the two Sierra provinces is concentrated in the regional centre and major city, Cuenca (Table 6.4). The city contains 110,000 or 72 per cent of the region's urban population (CREA, 1977, vol. 1, 40), and it monopolizes most of the region's commercial residential and factory growth. National government district offices are also concentrated in the centre of the city. For the other towns, including Cañar's provincial capital, Azogues, (11,000, 7.2 per cent of urban population) population growth rates have not been significantly different from those of the surrounding countryside. This is also true for most urban centres at parish level in the two Sierra provinces; in the mid-1970s, of the 95 cantonal and parish centres, 55 showed an increase in population and 40 have decreased (source: personal communication from CREA).

The provinces of Cañar and Azuay have traditionally been rural. This historical situation remains unchanged for Cañar, largely because the towns act only as

Table 6.3 Population growth, 1962–74

Province	*1962*	*1974*	*Difference*		*Net migration*
			No.	*%*	
Azuay	274,642	384,278	109,636	39.9	–47,673
Cañar	112,733	145,411	32,678	29.0	–18,137
Morona Santiago	25,503	53,312	27,809	109.0	10,557

Sources: III Censo de Población and information provided by officials of CREA.

Table 6.4 Urban population in Azuay and Cañar, 1962–74

City or town	*1962*	*1974*	*Difference*		*Annual % increase*
			No.	*%*	
Cuenca	60,817	104,470	43,653	71.8	4.81
Azogues	8,075	10,953	2,878	35.6	2.68
Cañar	4,958	6,727	1,769	35.7	2.69
Gualaceo	3,060	4,575	1,515	49.5	3.55
Girón	1,924	2,361	437	22.7	1.78
Biblián	1,756	2,141	385	21.9	1.73
Santa Isabel	1,641	2,068	427	26.0	2.03
Sigsig	1,234	2,021	787	63.8	4.38
Paute	1,515	1,998	483	31.9	2.43

Source: Personal information from officials of CREA.

periodic market centres and have few other economic attractions for the peasants and craftworkers of the countryside. In Azuay the situation has been altered by the accumulative role of Cuenca which continues to capture both labour and capital, and acts as a regional headquarters for the ruling groups.

Cuenca – a distant third among Ecuador's cities after Guayaquil and Quito, and showing a slower rate of growth than either – is the one focus of modern sector growth in the area. Even so, urban industrialization has not been sufficient in this regional city to attract large numbers of peasants and petty crafts people. The majority of the population – most of it Indian – still clings to the land in its villages or *caseríos* (clumps of individual huts and houses), and is only noticeable in the weekly *ferias*. But it is no longer isolated from modern influences. Rural people are aware of the changes brought by modernization and by the diffusion of new technology and modern consumption goods.

The city of Cuenca is the manifestation of this modern world just as, with its blue-domed and spired churches and impressive homes of its ruling class, it was always the expression of traditional power. But it is not the only, or perhaps even major, symbol of the new ways for the Sierra peasantry which now travels to the national centres of Quito, or Guayaquil, or to work in the commercial plantations of the coast. Cuenca is obviously the dominant pole of accumulation and the diffuser of modern lifestyles within its region. But as the centre of a poor rural periphery, it is subject to even stronger centripetal forces from Quito and Guayaquil than those it exercises over the three provinces (see Figure 6.2). It is a half-way stop in the process of capital accumulation. It creams off a limited amount of the region's surplus through primitive accumulation and unequal trade and prices, but even more surplus is channelled through it to the bigger, more profitable national centres; in turn, it acts as the regional recipient for the goods produced by the modern sectors of the major cities.

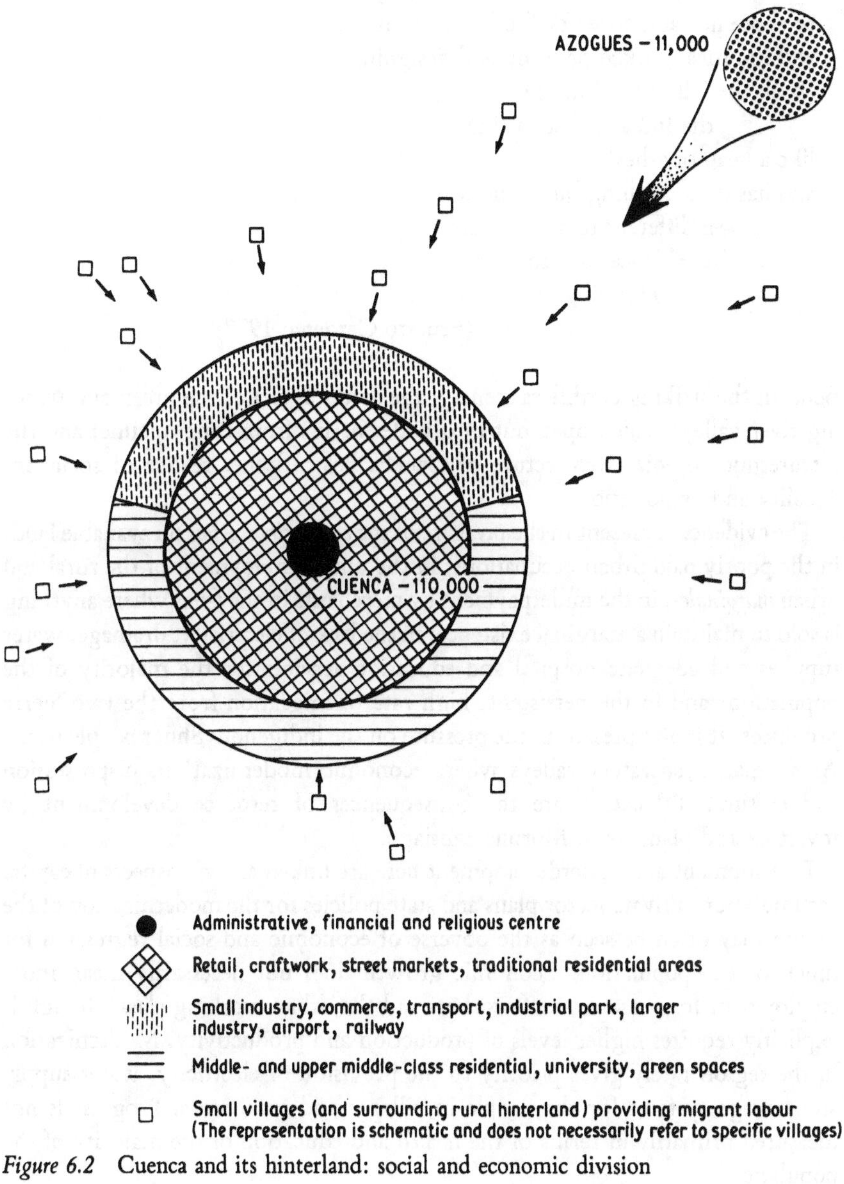

Figure 6.2 Cuenca and its hinterland: social and economic division

THE EVIDENCE OF INEQUALITY

TO KEEP THE INDIANS EMPLOYED
was an Inca slogan
the lame the disabled the old all worked
there were neither idle nor unemployed
to those not able to work food was given
and the Inca worked painting and designing
With the fall of the Empire
the Indian sat squatting
like a heap of ashes
and has done nothing but think . . .
indifferent to skyscrapers
to the Alliance for Progress
Thinking? Who knows?
(Ernesto Cardenal 1972)

Beneath the striking cordillera scenery, the softer aspect of the green and winding river valleys which open out into wider basins from time to time, and the picturesque colonial architecture of Cuenca lies a deeper reality of social inequality and exploitation.

The evidence is present in the pressure of the *campesinos indios* on available land; in the poorly paid urban occupations; in the underemployment of the rural and urban *marginales*; in the underpayment by merchants in the *ferias* where anything is sold to maintain a marginal existence; in the lack of electricity, drainage, water supplies and adequate hospital and education services for the majority of the population; and in the persistent, high rate of migration from the two Sierra provinces. It is also present in the pressure on the indigenous Shuar people of the Amazonian headwaters valleys where economic modernization, dispossession and cultural dislocation are the consequences of resource development by investors and planners in Morona Santiago.

Development and underdevelopment here are linked as twin aspects of capital accumulation. Private sector plans and state policies for the modernization of the region may often be seen as the obverse of economic and social regression for much of the population. Economic growth does not necessarily mean more employment for peasants, craftworkers and the urban working class, though it explicitly requires higher levels of production and productivity. Modernization in the region rarely gives priority to the provision of electricity, water supply and sewage systems for the people in villages and rural areas. Progress is not measured primarily in terms of the health and education of the majority of the population.

Development strategies, indeed, generally involve the transfer of surplus from the local peasantry, craft producers and urban workers to local and foreign investors. Furthermore, without changes in social relationships, any efforts which planners might make to develop the region will probably intensify the degree of inequality among the classes. Keith Griffin makes a parallel observation about the unequal relationships between the Sierra and the coastal region of Peru:

> Evidently the processes of development and underdevelopment are related; they are not independent of one another. It is the manner of integration and the resulting unequal distribution of benefits which in large part may explain the poverty of some nations and the development of others. The predominance of 'backwash effects' in some regions is closely associated with the degree of concentration of economic and political power. (Griffin 1969, 65)

Griffin is here concerned with the spatial or regional inequalities of a situation where capital accumulation in Lima results in the extraction of surplus from the Peruvian Sierra. 'It is quite clear', he says 'that internal migration, trade flows, and capital movements may have had absolute negative effects on the level of per capita consumption and the rate of growth of the poorer region' (Griffin 1969, 63).

But the relationship involves more than spatial links. It refers more broadly to the mechanisms that stimulate the expansion of the capitalist sector at the expense of non-capitalist activities; to the promotion of certain values, patterns of behaviour (consumerism) and cultural attitudes while undermining others; and above all, to the concentration of power in the hands of a ruling class. This provides a framework within which to consider the evidence on social inequality which follows.

The figures for education and literacy give an unduly rosy picture of a society where a person is considered literate by the signing of his or her name. But this does not take into account the fact that Spanish is an alien tongue to a great number of Sierra Indians who normally speak Quechua as their first language.

Table 6.5 shows that to be a woman or girl living in a country area is a guarantee of fewer educational opportunities. The rates of illiteracy do not vary markedly among the three provinces, though it is noticeable that Azuay, with a higher proportion of urban population, is comparatively better educated in all categories than either of the other two provinces. Rural areas and women are particularly disadvantaged.

Public funds designated for the provision of drinking water, electricity and sewage indicate the same rural/urban bias (Table 6.6). Furthermore, the largest expenditures for public facilities, even in the towns, have been those on roads and communications networks which link resource-rich areas, especially Morona Santiago, to Cuenca; these usually take precedence over the provision of access routes for rural populations.

Table 6.5 Illiterate population aged ten and above, by sex and area

Province	*Total population*			*Illiterate population*			*% illiterate*		
	Total	*Male*	*Female*	*Total*	*Male*	*Female*	*Total*	*Male*	*Female*
Azuay	266,648	119,088	147,560	61,180	17,220	43,960	22.9	17.2	29.8
Urban	88,673	38,932	49,741	6,807	1,322	5,485	7.7	3.4	11.0
Rural	177,975	80,156	97,819	54,373	15,898	38,475	30.6	19.8	39.3
Cañar	99,369	45,950	51,419	29,562	10,422	19,140	30.4	22.7	37.2
Urban	13,802	6,111	7,691	1,303	304	999	9.4	5.0	13.0
Rural	83,567	39,839	43,728	28,259	10,188	18,141	33.8	25.4	41.5
Morona Santiago	34,377	18,565	15,812	8,565	3,579	4,986	24.9	19.3	31.5
Urban	6,426	3,313	3,113	555	190	365	8.6	5.7	11.7
Rural	27,951	15,252	12,699	8,010	3,389	4,621	28.7	22.2	36.4

Source: Based on *III Censo de Población*, pp. 17–18.

Table 6.6 Water supply for private housing

Province	*Housing*	*Public system (inside/outside building)*	*Other sources (springs, rivers, tanks)*	*% lacking water supply*
Azuay	83,519	25,094	58,425	70.0
Urban	22,716	20,726	1,990	8.8
Rural	60,803	4,368	56,435	92.8
Cañar	30,737	6,060	24,677	80.3
Urban	3,657	3,318	339	9.3
Rural	27,080	2,742	24,338	89.9
Morona Santiago	10,047	1,266	8,781	87.4
Urban	1,798	888	910	50.6
Rural	8,249	378	7,871	95.4

Source: Based on *III Censo de Población*, pp. 56–67.

Statistics for Ecuador as a whole have been added (Table 6.7) to give a comparative idea of the region's standing, in a nation where provision of such public services is usually fragmentary. Only Azuay, presumably because of the city of Cuenca, comes anywhere near the national average. Here, even on a national basis, the regional centre appears to be in a favoured position, suggesting further discrimination between Cuenca and the rest of the region.

Evidence of economic inequality can be seen in land ownership. Table 6.1 compares rural population with the area of the two Sierra provinces, to give an average of 0.03 square km per person in Azuay and a little less in Cañar. But such a rough division of area by population fails to take into account the great variations in topography, soil fertility and availability of water in these mountain provinces – all of which considerably reduce the area of cultivable land. (The ratio of 90 per cent of population on 10 per cent of valley land gives a clearer impression.)

In fact, the distribution of the land is not known, as there is still no source of accurate data. The general impression in the region is that of fragmentation into *minifundios*, and, indeed, into *microfundios*, so tiny are some of the holdings. This situation is especially true near Cuenca and in southern Cañar, where, on hillsides irrigated by roughly constructed ditches and channels, peasant owners carefully tend miniscule segments of land measurable in square metres.

Yet, if the *latifundista* with his wide, fertile, underused hectares is not the same object of opposition as in other regions, there is evidence that inequalities exist in the distribution of the land, especially in the northern part of the province of Cañar,

Table 6.7 Housing and public facilities

	Housing total	*Electricity, not served*		*Toilets, not served*		*Drainage, not served*	
		No.	*%*	*No.*	*%*	*No.*	*%*
Azuay	83,517	55,462	66.4	61,286	73.4	61,652	73.8
Urban	22,717	1,323	5.8	3,568	15.7	3,499	15.4
Rural	60,800	54,139	89.0	57,718	94.9	58,153	95.6
Cañar	30,736	25,880	84.2	27,514	89.5	27,480	89.4
Urban	3,657	618	16.9	1,170	32.0	996	27.2
Rural	27,079	25,262	93.3	26,344	97.3	26,484	97.8
Morona Santiago	10,047	8,833	87.9	8,641	86.0	8,756	87.2
Urban	1,798	699	38.9	788	43.8	797	44.3
Rural	8,249	8,134	98.6	7,853	95.2	7,959	96.5
National	1,202,007	713,638	59.4	697,773	58.1	716,316	59.6
Urban	489,151	86,023	17.6	76,303	15.6	81,732	16.7
Rural	712,856	627,615	88.0	621,470	87.1	634,584	89.0

Source: Based on *III Censo de Población*, pp. 58–60.

and on the soils of the more fertile river basins. Absentee ownership, as we shall see, remains a reality and the old relationships of *huasipungo* still persist, despite modifications of such serfdom with the legislation of the 1960s.[2]

The influence of the old landowner class still persists in central Cañar, where just a few families have preferential access to the development loans of the government Banco Nacional de Fomento. Unfortunately for the province it seems that little of this has been intended for mechanizing and modernizing Sierra agriculture. Rather, it has been transferred directly to the landowners' export cash-crop plantations on the coast where the returns on such investment are higher (source: personal interview, Azogues). These same families also recruit labour for the coastal plantations. 'Se enganchan peones para trabajar en la costa', said one informant in the valley – 'Peons are roped in to work on the coast'.[3]

In this way, capital and labour for rural development are largely monopolized by the landowners. The more state investment loans that enter Canar, the more likely are the leakages of capital and labour from the capital-short province. The social implications, too, are of interest. The landowner with his formal feudal trappings in the Sierra is at the same time a capitalist export producer on the coast, and his peons in the Sierra are transformed into rural wage-earners on the coastal commercial estate.

The state, then, acts as subsidizer of the upper income groups. The lending policy of the Banco Nacional de Fomento tends to intensify the disparities between the wealthy and the poor in rural areas. According to Keith Griffin, (1974) in the mid-1970s the bank made loans to 6 per cent of all cultivators in any one year, but most were to large enterprises. About 65 per cent of the loans were for livestock enterprises and the average loan was close to $US3200.

The policy bias went still further. With nominal rates of interest at 8 or 9 per cent a year, and inflation rates of 20 to 24 per cent, credit recipients were benefitting 11 to 15 per cent. Public funds were, in effect, subsidizing large-scale farmers. On the other hand, *precaristas*, *parcelistas* and *minifundistas* had recourse mainly to non-institutional lines of credit, money-lenders and traders. Borrowing until the harvest, they paid high rates, set monthly or even weekly, to cover inflation and usurers' risks. Cheap institutional credit, as Griffin notes, also affects the structure of rural production; it tends to encourage investment in capital-intensive technology which, in turn, is likely to decrease employment opportunities in agriculture (Griffin 1974).[4]

Local school teachers and others with knowledge of the area, speak of the historical roots of poverty in the north of Cañar. Property has been concentrated in the hands of *Cuencanos*, members of traditional land-owning families now living in the city of Cuenca, whose principal interest in the region has been to extract surplus from their estates to support a comfortable urban lifestyle.[5]

Between these two traditional extremes of large-scale absentee ownership and the intensive production of the fragmentary peasant plots, a new class is emerging in the countryside and small villages. It is a category which is now no longer *minifundista*, but has not reached the level of large owner. From the peasantry is emerging a group of small-owners who, through good fortune, more fertile land, better management or swifter adaptation to modern circumstances, has raised its condition above that of its neighbours. The relatively comfortable position of these farmers gives them an advantage which tends to increase cumulatively, so that in times of hardship they have the reserve capacity to survive economic difficulties. This, in turn, may enable them to become their neighbours' creditors. Of even greater significance in the process of capital accumulation, they have the capacity to amass a surplus which allows them to rent or buy the land of others and to employ the labour of their neighbours.

This is not an unusual phenomenon in history. Maurice Dobb, dealing with the transformation of the peasantry into the proletariat, discusses the process of enclosure and forced eviction by the ruling classes, as one way of accomplishing the change. But there is also another way:

> It consists of the tendency to economic differentiation which exists within most communities of small producers unless special institutions prevail which are capable of preventing inequality. The chief factors in this differentiation are differences that arise in course of time in the quality or quantity of land-holding

> and differences in instruments of tillage and of draught animals; and the agency of eventual dispossession is debt. (Dobb 1963, 242)

He quotes a number of historical examples, but perhaps the most relevant one is to be found in Lenin's *Development of Capitalism in Russia*; by the end of the nineteenth century, the traditional mir or village commune

> was destined to disintegrate in the face of the influences of the market and was already well advanced on the road towards capitalist agriculture, with the growth of class differentiation among the peasantry. (Dobb 1963, 251)

Dobb's observations by no means imply a simple or linear progression in social evolution; nor (as we indicate in the case studies) are they precisely applicable to the Ecuadorian Sierra. Nevertheless some elements of the historical examples can also be seen at work in the changes taking place in the region.

THE EVOLUTION OF A RURAL COMMUNITY

The first case study community is located in a gradually narrowing and winding valley in central Cañar, whose rolling corn-planted fields are studded with the adobe and straw-roofed *chozas* of a closely settled peasant population. The density increases until the houses run together to form a village settlement of perhaps two dozen buildings, a primary school, a shop and a new church, built along three main unpaved streets. The village seems more an intensification of the general pattern, hardly different in kind from the close settlement of the rural part of the valley. Every piece of land is in use for the cultivation of crops or, on the steeper slopes, to graze goats, donkeys or pigs. The careful management of water is the key to subsistence in the valley. It is channelled away from the valley's main watercourse, past the *chozas* around hill slopes and down roadside gutters in a myriad of canals, and trickling streams, dyked and dammed to feed fields of corn or tiny beds of lettuce, potatoes, beans and cabbages for the weekly provincial market.

As the deeply rutted earth road runs further up the valley into the deforested and eroded hills, the valley sides become steeper and the pressure by peasant cultivators on the narrow, irrigated defile becomes even greater than on the easier slopes of the lower valley. Here it is not just the agriculture of the *minifundio*, but of the *microfundio*.

Throughout the valley as a whole, the lots have been small, sometimes measuring only a few square metres. But the *minifundista* knows to the last centimetre the subdivisions of the vegetable patches, marked by borders consisting of flowers (chosen for their marketability), rows of vegetables, narrow paths or irrigation channels. A few lots have been abandoned and some are poorly cultivated, but the great majority are worked up to the edge of every boundary. The general rule is that, on the small lots, the owner concentrates intensively on vegetables, but where there is more land it is possible to cultivate *maiz*, the food staple of the Sierra. Only on parcels of land of one-quarter hectare or more is this possible.

This pattern is taken to its extreme in the upper part of the valley. The plots become mere patches of ground, of perhaps two metres square, crowded with carrots, beets and beans. Others are nothing more than strips of land between rows of corn, planted in potatoes and other vegetables. It is, to the outsider, an unbelievably confused mosaic of ownership, but not to the owners themselves or to their neighbours with their own dwarf pieces of land.

Probably it will not continue in the future. But for the moment, the people cling obstinately to their distinct bits and pieces, distributed widely throughout the valley. It is *their* land; it represents their security and place in the valley and community. Their situation is marginal in many cases. But for the moment, they may be able to stave off the worst of the economic pressures, even if at social cost to the family and the community.

The adaptability and ingenuity of the Sierra Indian peasant family certainly arise from the threat of poverty, but also find their roots in a cultural commitment to the land and the strength of community and kinship bonds. To the agricultural product, part for subsistence, part for sale at the weekly market in Azogues, can be added the returns from household craft production. The valley is one of the areas making the famous 'Panama' hats, fabricated from the special straw, *paja toquilla*, which is brought into the province to be woven by the skilled fingers of the Cañari women. Other villages in the region specialize in leather, wood, metal and silver products as well as making clothing on a putting-out system, either for the traditional market or for the factories and merchants of the middle class.

Differentiation among families is now taking place in the community. Don Fernando is a landowner who is establishing a superior position in the valley. He owns a small, old truck and the only all-service store, with its limited selection of modern consumer goods, in the village. He sells liquor, especially the local *aguardiente* distilled from sugar cane, some factory canned goods, but little local produce, and *colas* bottled in the city. His store is the meeting place for local men, and in discussion he himself is usually at the centre of the group. The signs of restrained deference he receives derive also from his position as an employer of smaller peasants on his half-hectare *potreros* of land.

The relationships between Don Fernando and those who work for him, weeding his carrots, potatoes, corn and cauliflowers, are not those of an owner with wage labourers. They are more subtle and complex, the links of those who are all proprietors of land, members of the same community, though of varying fortunes. In his fields Don Fernando is obliged to participate occasionally in the hoeing or weeding, and from time to time to pass around the glass filled with the fiery *aguardiente*, 'para la fuerza', to give strength. As he explains, his workers might be prepared to work all day almost without pay, but not without their *trago de aguardiente*, and a demonstration of effort from their *patrón*-neighbour. He is in an intermediate, transitional position, a wealthy neighbour and employer, a man of more than ordinary importance, a leader, a purveyor of commodities, transporter of goods to and from the market, and, in some cases, commercial intermediary in the *feria* (market) of Azogues.

Such relationships seem to contain elements similar to the evaluation described by Dobb, but the comparison should not be pushed too far. It is much less likely that the landless peasants in the valley will shortly evolve into urban proletarians as their European counterparts have. No jobs for an incipient industrial proletariat await the peasants of Canar in the provincial capital of Azogues; there is virtually no industry there apart from the same sort of artisan craftwork they practise in the countryside. The town, in fact, comes to life on Saturdays when its population trebles or quadruples and the streets and market squares fill with its crush of mainly rural buyers and sellers.

The significance of the feria

Throughout the three provinces, the *feria* is the principal social and economic link between the towns and the countryside. The weekly gathering may appear to the casual onlooker to be the occasion for the fortuitous interchange of traditional products – the remnant of an outdated and passing system of exchange. Such is far from the reality; it is not a chance affair for the buyers and sellers in the bustling, crowded streets. Each of the different products has its specified area – animals down near the river, fruit, corn and vegetables near the central square, fabrics and clothing on the other side of the square, small metal objects, pottery, some stolen goods in a side street, coke and firewood in the upper part of town.

Many of the products are traditional, grown and fabricated in the small villages and rural areas. But modern, urban consumer products including cheap mass-produced plastic goods and factory-made metalware (aluminium bowls and saucepans) from the cities are now appearing alongside their craft equivalents, in increasing numbers. Changes in clothing styles especially for men (though less for women) and the attraction of transistor radios and wrist watches among the wealthier peasants, are all aspects of a modernization process stimulating new consumption patterns as well as the felt need for status and achievement.

The market is also a means of surplus appropriation by local merchants and those from the larger cities. In its use of the *feria*, merchant capital acts as an intermediary, bringing factory goods from the big cities to the smaller towns and from one region to another. On every transaction it takes its profit, the size of which depends on the level of knowledge and distance between the local buyers and sellers and on its control over producers. The traditional market is thus of great value to merchant capital, which profits through unequal trading and pricing practices.[6]

The larger merchants move among the different town markets from the regional centre of Cuenca, buying wholesale there in the Thursday market, to sell retail in Azogues (or other market centres) on Saturday. Small traders, too, compete over the short distance between the two centres, but are at a disadvantage over longer distances. Within the Azogues market, small vendors, in turn, buy wholesale and sell retail at various times during the day of trading. The trickles and rivulets making up the cash flow from producer-trader to merchant – the basis for the centralizing process of capital accumulation – are thus constantly taking place through adapted traditional institutions.

Much has been written on the midwife role of merchant capital in helping the birth of industrial capitalism, since Marx devoted part of *Das Kapital* to the question. Dobb (1963) and Kay (1975) provide insights into the contradictory role of mercantile capital in both promoting and, at the same time, hindering the evolution of industrial capitalism, while Latin American writers have used the question to throw light on the region's particular problems within the world capitalist system.

Theotonio dos Santos observes that, while commercial capital is a necessary condition for the emergence of industrial capitalism, it is not a sufficient one. The operations of such capital can even block the birth of capitalist production under certain circumstances. Not only, he argues, is it wrong to assume that it led historically to the development of industrial capitalism in Latin America:

> As well, we might explain a large part of our underdevelopment by the great predominance and development of commercial capital in the colonies, a factor which limits the creation of a capitalist mode of production. . . . Production has been born among us fundamentally as the production of merchandise, but it is a production retarded and subordinated to commercial capital.
>
> (dos Santos 1973b, 123)

Dos Santos is writing here of a continental situation, but the argument can also be applied to the three provinces, as we try to show in the following chapter.

Migration – a rural response

Pressures on the land and surplus appropriation lead also to the movement of rural people. The increasing centralization of capital in the large urban centres is paralleled by a migration of labour in the same direction. For some the move is permanent, for others – those determined to hold on to their rural roots – there is a series of departures and returns to origins, repeated on a monthly, weekly, or daily basis.[7]

The reasons for such movements are set out in a University of Cuenca publication:

> In effect, in a country where the majority of the population lives in rural zones and carries out agricultural activities, the rigid agrarian structure characterized by the existence of the *latifundio-minifundio* complex constantly expels labour to cities which have reached a certain level of capitalist development.
>
> Thus, in those regions where *haciendas* are dominant, the introduction of capitalist forms of production – implying a rationalization and mechanization of work – displaces workers who, finding it impossible to obtain any other occupation in the rural zones, must then migrate to the cities.
>
> On the other hand, the *minifundio* structure which is dominant in this country, is the principal reason for this migratory current, either because of dynamic population increase, or through a decline in production and productivity, as a result of the impoverishment of the soil. It expels the younger people from the rural zones, and produces a veritable invasion of peasants searching for some form of work in the cities. (Bravo, n.d., 2–3)

Where, specifically, are the rural migrants headed? Very few look to Azogues, provincial capital though it is, or to the smaller cantonal towns. Such centres are qualitatively not very different from their surrounding rural areas. Except for some minor activities, their employment structures have limited absorptive capacity. They are organic extensions of rural life, its peasant and craftwork systems and its culture, while also acting as minor centres of primary accumulation for trading capital. What they do not offer, however, is an escape valve of alternative or additional employment for the rural population.

Population changes in rural communities

Some specific case studies from Sierra villages will help to illustrate the migration process. The range of solutions among the villages and parishes of Azuay and Cañar described here is further evidence of the ingenuity and adaptability of a

Sierra peasant and craftworker population under pressure. It is a measure, too, of the determination of rural people to hold on to the security of their land, however tiny the plots, and to survive in a rapidly changing world.

Suscal in the north of Cañar is a small village of nearly 600 people encircled by mountains and the mist of their clouds; it lies just off the gravel road forming a large part of the road link between Cuenca and Ecuador's largest city, Guayaquil. The unadorned central square of earth and stones, lined by low wooden buildings – only the church and its school alongside are of concrete – gives the impression of unchanging solidity. It is a false impression, for the population is in a state of constant flux. Some of the children playing in the square were a short time before living with their families on *parcelas* in the nearby rural parishes, while others may be on the point of leaving to seek work in Cuenca.

According to the school director, the situation in the village and in the surrounding countryside is fluid. Peasants move in to the village to take over the homes or shops vacated by those who are leaving for Cuenca or the coast. Other peasants – the area is principally a producer of corn and wheat – leave their parcels of hillside land as often as three times a year to work as wage-labourers on the export plantations of the coastal province of Guayas. Their employment is seasonal, and they themselves represent the type of 'free' labour for which the employer has little responsibility. They are a reserve workforce for the commercial plantations, available for the planting of the crop, later for the weeding, and finally for the harvesting.

The modern agricultural sector thus maintains itself at least cost: a plentiful supply of labour just when the employer needs it keeps wages low, and at the same time there is no obligation for him to provide social welfare facilities for the sick, the uneducated or the unemployed. Three interrelated levels of surplus appropriation can be detected here: that of the coastal region which is subsidized by the Sierra; the capitalist sector which draws upon subordinate systems; and finally, the commercial farmers who employ migrant Sierra peasants and craftworkers as well as their own agricultural labourers.

Yet the balance is not entirely lopsided. Such constant movement in recent years has had its impact on Sierra peasants, for long bound to their own *parcelas* or to the lands of their *patrón*. They are now 'mas despiertos', more awake, to alternatives and new opportunities. The transistor radios they buy have widened their world. And they have lost a certain amount of awe for the church, for centuries a dominant force in their daily lives. In the case of Suscal, this may be reflected in the fact that the local priest, for lack of funds (tithes, offerings and gifts), must now teach in the local school.

In El Tambo, a village on the main highway to the south of Suscal, the local school director described a similar situation. The population at the 1974 census (1375), had increased only by 0.99 per cent a year since the 1962 census; this, given the high rural birth rates, denotes a constant outflow of people. Some redistribution of land took place in this northern canton of Cañar but usually in tiny grants, and *latifundios*, considered large at 30 to 50 hectares, still occupy the fertile river valley lands. Meanwhile, on some absentee-owned land, local leaders of rural Indian communities and progressive priests have formed groups of peasants to distribute land and establish agricultural co-operatives after waiting in vain for action by the state agrarian reform agency, IERAC, or the regional development agency, CREA.

Nevertheless, people are moving out of northern Cañar, bypassing Azogues, their provincial capital, and heading to Cuenca, often to settle there permanently. And, as with the migrants from Suscal, those with a little capital take it with them to the city. The other movement is the seasonal one to the coast, to work in the large sugar, rice and banana plantations before returning to rejoin families engaged in agriculture and craftwork.

Official programmes have been proposed – improvements in education in the technical college, established by the previous military government, and the creation of agricultural processing industries are two possibilities – but little local confidence was shown in such offers. Despite long delays, there were still no ministerial funds to provide teachers for the special technical and agricultural courses in the government technical college. Nor was the situation very different in the eight other special technical colleges in Azuay and Cañar, except in one school operating on the principle of self-help.

The third example of a people's adaptation to adverse circumstances is to be found in a long narrow valley in the southwest of Azuay province. The settlement of Gima stands where the dirt and gravel road, following a narrow winding defile down from the main Sierra highway south, opens out onto a widening valley. This becomes a more spacious undulating plain stretching eastwards across to the cordillera and the high *páramo* which divide the Sierra from the valley of the Upano river and the Amazonian region of Ecuador.

The principal commercial activity in the valley is concerned with dairy production, and the village itself boasts a small cheese factory behind the square. There appears to be a higher-than-average level of prosperity, based on larger farm size and the production and sale of cheese, which is something of a luxury good, and this holds true for some of the valley's inhabitants. A small group holds properties of up to 30 hectares on the more fertile undulating lowlands – virtual *haciendas* in this province of miniscule peasant holdings – but these proprietors are a minority. Most of those who can claim a piece of land are peasants with subsistence parcels as small as a half hectare; one informant in the village thought that the average size of holding for the valley might be about 5 hectares. Another, who had left the valley to teach elsewhere, made the more significant point that polarization of property ownership was already occurring;

at the other end of the scale from the many patches of *minifundio* land there was already in existence a 180-hectare *hacienda*.

Again, such social differentation within the valley community manifests itself most clearly in the search for work. Many of the *minifundistas* and their families cannot subsist on their tiny plots. A few manage to find part-time local employment, working for others on farms, in shops, or in the factory, but such outlets are inadequate for the majority of the people in the valley. Each year, therefore, two different groups in the community follow their particular migration routes.

The first is taken by the *emprendedores*, the more 'enterprising', those with larger amounts of land who migrate seasonally to the Oriente. There, down through mountain passes to the province of Morona Santiago, they pasture their cattle on their *fincas de colonización*, their spontaneously colonized farms, using the *granmalote* and elephant grasses as well as the natural pastures which have emerged in the cleared areas. To round out their activities, they plant sugar cane, *papachina* and *yucca* on *fincas* of up to 40 hectares, and distil cane sugar into the regional beverage, *aguardiente*.

The small peasants, lacking opportunities for acquiring Oriente lands, with inadequate *parcelas* in the valley and with limited local employment opportunities, must migrate seasonally to the coastal province of El Oro. There they compete to provide the same form of cheap wage-labour on the export crop plantations as their fellow *serranos* from northern Cañar do on commercial estates in the province of Guayas.

The reasons for unemployment and underemployment may vary in detail among the villages of the Sierra provinces, but the resulting pressures and forced solutions are comparable for those drawn into the orbit of the modern capitalist sector. Most income groups are affected economically by the modernization taking place in the Sierra communities. The social, political and cultural consequences are becoming evident, although the surface impressions of the society often appear unchanged.

The experience of a number of villages further north in the region, however, gives an additional perspective. It tells, in part, of the determined effort by the villagers to hold on to their land and to their community and kinship bonds by remarkable adaptation to both the pressures and opportunities created by modernization. In two of the three villages, the 1974 census revealed only small increases in population since 1962 – barely one-third of the national average – while the third recorded incredibly high rates of decrease for a functioning, live community. There was also a disproportion between the sexes, with women far outnumbering men among the adult population.

The causes lie partly in the seasonal migration to the province of Guayas on the coast, where the heads of families and older sons work on the banana, rice and sugar estates, while others leave to try to find urban work, usually in the petty trading sector of Guayaquil. Many among the latter then send for their families, who make their way down to the city by bus or truck with their belongings, swelling the ever-expanding circle of *suburbios* ringing the city. The thousands of stilted, cane-sided, straw-roofed shacks sticking out of the mud of the tidal estuaries around the port city provide the most telling evidence in Ecuador of the Sierra provinces' problems – and of the inadequacy of regional and social policies.

Few move from these villages to work in Cuenca or Azogues. Bus services – two or three a day – are used by the people mainly to carry their goods to and from the market. But the villagers, apart from a few young people, have found few jobs in either city.

The feature which gives these villages their unique character in the region is the unlikely destination of one migrant stream. By various means a channel has been opened for migration to the United States. The clandestine operation has required the mediation of a middle-class *emprendedor* in Cuenca who, for a certain consideration, obtains the necessary travel documents and visas for those travelling.

And so the chosen heads of families are sent to work in the United States, in poor working conditions and at low rates of pay – up to $390 monthly in the mid-1970s was a figure mentioned by one informant – and without recourse against unscrupulous employers. Nevertheless, this represents an opportunity to earn sums unmatched at home. Wages are sent home by post, a practice which, from time to time, allows for 'intervention' by postal officials, so that, in fact, the migrant suffers a double exploitation. Even so, the inflow of dollars has allowed certain families to improve their material living conditions, and they have invested in consumer goods, farm machinery, trucks and property in Cuenca. The properties are then rented out, and the returns reinvested in the urban centre rather than the villages.

Here we have an illustration, extreme perhaps, of the changes taking place in what has for centuries been a conservative, Church- and landowner-dominated, traditional rural society. Class differentiation within the village community; the channelling of capital into the urban centres; the changing attitudes of those who have travelled and experienced capitalist modernization; and the effects on those at home who have become observers of the consumer society – all these currents, moving beneath the observable surface of Sierra society, provide alternative perspectives, social alignments and attitudes towards previously accepted customs and institutions.[7]

But this does not mean that there is an inevitable move towards defined alternatives by an increasingly aware majority. The flexibility, the mobility and the act of migration itself are responses to the need to find solutions to individual or family problems as well as the wish to avoid confronting a sytem which creates such problems. They are the non-violent escape valves used to avert conflict with an authoritarian ruling class and to allow families and communities to hold on to the bases of their daily lives.

The discussion on the migration of the Sierra people has so far concentrated principally on the seasonal movements of peasants and small-village dwellers to the capitalist export-crop sector of the coast. We have mentioned, too, the more permanent move undertaken by many who are drawn to, and settle in, the spreading *suburbios marginales* ringing Guayaquil. A similar, though less numerous, flow of migrants north, up the spine of the Andean cordillera to Quito, the national capital, has also been characteristic of outmigration from Cañar and Azuay. The oil boom of the 1970s which transformed the face of Quito and brought wealth to the city's bourgeoisie has now made it attractive to the *marginados* seeking some spill-over from the wealth accumulated in the nation's capital.

Cuenca: regional centre

But some rural migrants from Cañar and Azuay have also found work in Cuenca, where capitalist modernization of production and consumption is most in evidence. The city also exercises a more immediate and direct impact on this mainly rural society. Its socio-economic influence stretches out some 15 kilometres to encompass a mosaic of small rural villages within its radius of direct influence (see Figure 6.3). Daily, dozens of buses make numerous trips between the city and its satellite villages bringing people to work, mainly in artisan production and the small-time trading and informal service sectors.

There is also another group which is in a different category: the wage workers in Cuenca's small and slowly expanding modern capitalist production sector. They are not a large group, because Cuenca is a secondary city whose economic structure is based more on commercial than modern factory activity. By definition, they should be clearly distinguishable from those working in other modes of production, but in practice the different groups of workers cannot be so easily distinguished as belonging to the industrial proletariat, or the peasantry or craft occupations. The social boundaries are smudged, the activities overlapping, as indeed the differences between town and countryside are blurred.

And Cuenca itself, because of its late arrival as a centre of capitalist production (though much earlier as a diffuser of modern consumption values), is a city *sui generis*. Its morphology and built environment are still those of a colonial, religious, administrative and market centre, laid out on a grid pattern which is defined and visibly clear cut. Only now are the squared-off boundaries of the city's colonial limits becoming rounded out, or frayed at the edges: recent commercial development now pushes out along the highway to the north leading to

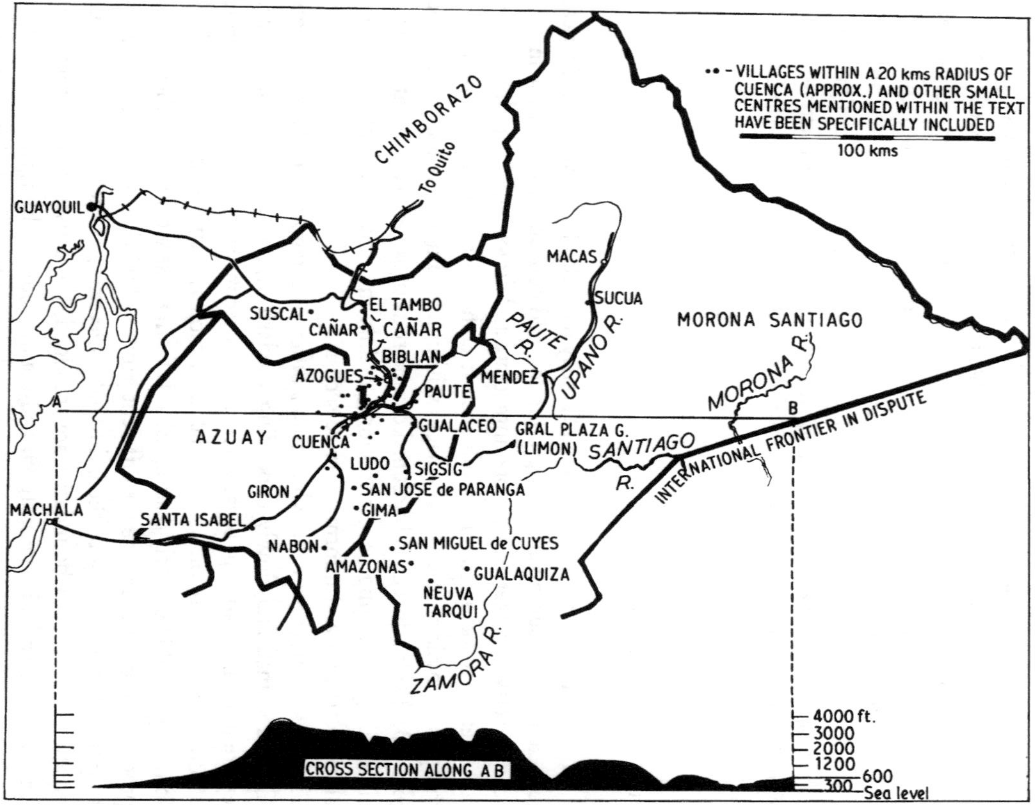

Figure 6.3 Azuay, Cañar and Morona Santiago, showing villages within a 20-km radius of Cuenca

the airport, and to the hopefully waiting, but still largely empty, municipal industrial park. As befits a city which is modernizing, new villas and chalets are springing up on the outskirts of the old city in wealthy suburbs, which proclaim in their consolidated wealth the benefits, for some, of capital accumulation.

The centre of the historic city remains, outwardly at least, little altered. The streets, now converted into one-way thoroughfares, run at right angles to each other and incorporate market squares, each specialized in its commodities and trade. At the centre stands the Plaza de Armas, a tree-filled square overlooked by the blue-tiled cupolas and rose-marble slabs of the cathedral. Cuenca still has the feel of a colonial city; the narrow, cobbled streets, the long, flat adobe frontages of the two- or three-storey buildings lining them, and the low sweep of urban skyline broken by the spires and cupolas of the dozen or more churches all proclaim its history.

Under this surface, change is taking place. Many of the great homes of the traditional bourgeoisie, once opulent and spacious behind the flat adobe walls, are now divided up and filled to the last square metre of space by the migrants from the region's villages and countryside. The newcomers to the city, as in the *tugurios* of Quito, have hidden their hardship behind the white walls. Poverty and exploitation here, as in the bustling markets, or among the women washing clothes in the Tomebamba River, become threads in a picturesque fabric which captivates the tourist looking for local colour and traditions. The blemishes which extend through the great arcs of misery in the *barriadas* of Lima, the *suburbios* of Guayaquil, the *campamentos* of Santiago hardly deface Cuenca yet.

Cuenca has its bootblacks, domestic servants, washerwomen, street sellers and carriers of goods, its thugs, beggars and prostitutes. But the problem of marginality has its own peculiarities here. The apparent absence of concentrated clusters of misery in *suburbios* that are so highly visible elsewhere on the continent does not, unfortunately, dispose of the *marginados*; as a local writer notes:

> they exist, and in a great quantity, but these masses are not found located geographically in one determined part or suburb; rather, their housing is spread throughout the whole city, and especially in nearby peasant localities. (Bravo, n.d., 5)

The marginalized of Cuenca thus comprise two groups: those living within the city limits and those from the surrounding villages and rural areas. The former are important in having changed Cuenca socially and spatially by replacing the wealthy, who have moved from the noise and crowded conditions of the city centre to more commodious suburbs three or four kilometres from the commercial and administrative district. The latter group, however, provide the majority of *marginados* who still live in the many small satellite communities within 15 kilometres of Cuenca.

> This allows them to combine their urban work with agricultural and craft-work activities which are carried on in the rural parishes. (Bravo, n.d., 6)

The conclusions Bravo draws from this are not particularly optimistic for the poor, although they may offer a sense of confidence to the ruling groups that the *status quo* is not likely to change in the short term. In the first place, the dispersion of the marginal groups decreases the chance that they will form pressure groups to demand concessions. Since they hold tenaciously to their close ties with rural life, they also retain the attitudes of peasant producers, craftworkers and street vendors – above all, an individualism which may leave them more open to manipulation by the ruling groups.

Moreover, the very flexibility of those who are peasants, craftspeople and proletarians at one and the same time, works in favour of their greater exploitation.

> It is this very phenomenon which allows us to explain the low salaries which are paid in this city (in the mid-1970s about S/.30 a day). It is below the minimum subsistence level, but is complemented by the economic assistance which derives from agricultural or craft activities on which he and his family work in their free time. . . . It is this violent form of exploiting labour which explains the rapid accumulation of capital by the dominant classes of Azuay; that is, we can observe a process based on the superexploitation of the labour force, in which capitalist production uses the precapitalist forms still existent in the society to lower the cost of maintaining and reproducing labour-power.
> (Bravo, n.d., 4)

Bravo discusses further characteristics of Cuenca's modernization. The reasons for the particular patterns of development in the city result, in part, from its relatively recent modernization. The history of import substitution industrialization in Ecuador compared with Argentina, Brazil or Chile, is very recent, even in the national centres, Guayaquil and Quito. For the distantly third-placed city of Cuenca, the experience has hardly begun.

Moreover, it seems to fit the pattern (outlined in Chapter 4) of many regional cities whose primary role is to act as mobilizers of the flow of accumulated capital to the larger cities. Cuenca is a commercial intermediary between its own hinterland and Ecuador's principal cities as, in turn, are Quito and Guayaquil with respect to world centres of accumulation and investment.[8]

Hampered by its weak, lower-order urban position, Cuenca has a limited capacity to attract either labour or investment capital. While Guayaquil can exert an attraction throughout a whole nation – and has recently been joined by Quito in this capacity – Cuenca draws relatively few migrants from the more distant parts of its region; these are some of the people who are now filling the old quarters of the city core. The more significant labour component is drawn from within a small radius of Cuenca, taking in the villages and densely populated countryside of the *minifundio*. These are the workers who link city and countryside, and whose low-cost labour provides the basis for much of the urban accumulation.

The uneven and partial nature of this capitalist expansion has created only limited possibilities for employment in the modern commercial and industrial structure of the city. In turn, the narrow base of the modern sector has limited the growth of the urban working class, and it has forced the migrants to seek work in the non-capitalist sectors. The peasantry/craftworker/wage-labour population is not entirely rural or urban; it is often reliant upon all three forms of activity to maintain itself, and dependent upon the urban employer and a labour-surplus market which determines and depresses the level of wages. In such a situation, this population-in-transition is weak and divided; and the consequence, according to Bravo, has been 'superexploitation'.

The beneficiaries are the employers. As Meillasoux has pointed out, the countryside, the small satellite villages and their communities provide a 'fall-back' for the sick, the old, the young, and unemployed and underemployed. In the absence of social welfare facilities provided by the wider society, the rural community continues to act as the socio-economic buffer for its inhabitants, and subsidizer of costs for the capitalist sector. In turn, rapid population growth can be seen as 'the logical means to face the social security requirement' (Meillasoux 1972, 103).

The structures of inequality begin with primary accumulation – the many runnels and rivulets of capital flow from countryside and village peasant/craft production to market towns and thence to the larger urban centres. These flows consist of unpaid and underpaid labour and services, together with taxation, the tapping of local savings by the modern business sector, discriminatory pricing systems, unequal interest payments and debt.

Part of the process also involves the modern sector and the direct exploitation of wage-earners caught in the transition to modern factory production. Bravo, in his report on 2500 construction workers in Cuenca, notes that although a minimum salary of S/.1250 (sucres) existed in 1976, 77 per cent of workers in private firms received less than that floor. Treatment was somewhat better in the public sector with 11 per cent taking home less than the minimum. On a six-day week, the minimum at that time gave a daily wage of S/.52, or rather less than $US2.

Many, in fact, were receiving somewhere between S/.30 and S/.40 ($US1.00 and $1.50) a day for a total of between S/.720 and S/.960 ($24 to $36) a month.

But Bravo discovered that for many families, with father, mother and some of the children working, the wage per person was still lower. Two-thirds of the families interviewed were receiving monthly incomes of 300 sucres or less per person which created a socio-economic situation of real anguish.

> The data demonstrate . . . the degree of exploitation imposed by the construction companies on their workers. This is especially true of the private firms, and indicates to us how the expansion of the Cuenca bourgeoisie is the product of a pitiless superexploitation of labour. (Bravo, n.d., 31)

Nor did the construction workers form a special case. In a modern ceramics factory, foremen or lead hands received S/.80 (less than $3.00) daily (S/.1920 a month) and ordinary labourers S/.50 (less than $2.00) daily (S/.1200 monthly).

Moreover, the income differences between those in the capitalist sector and others in petty production, trading or services were not always significant. A rough calculation of the earnings of a shoe cleaner in the Plaza de Armas, working in a favourable location next to the cathedral, helps to confirm this. At S/.2 for twenty to thirty clients a day, earnings could reach S/.50 or even S/.60 a day. Working seven days a week he could bring in S/.400, and with luck, up to S/.1600 a month. A more realistic figure, though, would be closer to the legal monthly minimum of S/.1250; and, for those in not so favoured locations as the central plaza, considerably less. Under such conditions, it would be difficult to speak of a 'labour aristocracy' in the modern sector of Cuenca.

In the transitional phase through which the city and its rural surroundings are passing, the range of occupations and activities is widespread and complex. *Trabajo casero* or cottage industry illustrates the ability of capitalism to use all the possibilities for reducing costs. The putting-out system employs women in their *chozas* to make garments for the market, so allowing the employer to avoid many of the final costs of production. It is a system with many precedents in other transitional societies, both contemporary and historical.

Maurice Dobb has described the process in eighteenth-century England in which the persistence of primitive types of production delayed the development of industrial capitalism. When the rural semi-proletariat was finally uprooted and forced into the cities, capitalism moved more rapidly towards maturity (Dobb 1963, 230–1). In the meantime, such practices delayed the evolution of the urban working class as an organized social force and allowed capital to augment its control over labour (Dobb 1963, 265–6). The circumstances are different, but there are elements in Dobb's analysis to compare with the changing situation, as Bravo outlines it, for the construction workers of Cuenca – and by implication for all those caught in the interstices of an evolving society (Bravo, n.d., 66–7).

The uncertain and fragmented nature of the work limits possibilities for labour organization. Eighty per cent of construction workers have never been members of unions (this rises to 90 per cent in private sector firms), while 93 per cent have never participated in strike action. Throughout the manufacturing sector as a whole a significant minority remains outside any workers' organization, as a report on trade unions in Azuay confirms:

> A third of the organized workers in the secondary sector who can be considered strictly as proletarians in the province are not integrated definitively into the labour movement and the percentage would be greater if all those in the sector were taken into account. (Carrasco, n.d., 114)

Bravo describes the effects of social fragmentation and confusion of this period of transition. The peasant and artisan, partly uprooted from the land, yet clinging to it, are subjected to all the strains of living in two worlds. Despite resistance, the patterns of material life and social values are in the process of disintegration under the challenge of capitalism's expansion. The individualism and the struggle to retain a hold on the land remain, but they are constantly jolted and shaken by the demands of urban-centred modernization.

> However much these workers might maintain a permanent tie with their rural past, the fact of the daily trip to the city and of being in direct contact with urban life, obliges them to adapt to this new reality, its values, its institutions, and its practices. But we must also not forget that, by not completely breaking their contact with rural life, the value patterns, customs, and the mystic mentality of the peasantry persist, so that their vision of the world reflects a contradictory and ambivalent coexistence. (Bravo, n.d., 72–3)

His conclusion deals with the social and psychological results of this evolution:

> This dualism in values, we believe, provokes contradictory situations of great tension in the mental structures of these semi-migrants; these show up in behaviour in which resigned and submissive attitudes of rebellion and of violence, almost always irrational, are expressed against a system which imposes on them brutal forms of exploitation, and, further, leads to a type of identity crisis. It is also one more factor pushing these peasant migrants to seek refuge in certain forms of social escapism, such as religion and alcoholism. (Bravo, n.d., 73)

Only occasionally have there been organized outbursts of protest against the ruling groups and systematized exploitation, when street manifestations give a certain brief focus to mass discontent. A more significant indication of the externalizing of anger and resentment by the exploited may be the rise of delinquency and violence in the city. It may be counted as another measure of the city's advance toward modernization.[10]

We have dealt here with one side of the changes taking place in southern Ecuador under the pressure of modernization. The extraction of surplus is the primary act; the second step concerns the uses to which surplus, as accumulated capital, is put by the appropriators. They are the subject of the next chapter.

7 Ecuador: capital and the accumulators

Introduction

We come now to consider the other elements in the process of capital accumulation: those who accumulate, the methods by which they appropriate surplus, and the ends of appropriation/accumulation. The ruling groups of Cuenca and their activities there and in the larger urban centres, the Church, and the state and its policies will be the principal characters in this enquiry.

It is difficult, if not impossible, to generalize about the ways in which local ruling groups establish and maintain their power. This is especially true under conditions of capitalist modernization where the development process assumes different forms and pace in each of the areas it touches. Walton's (1977) studies of four Latin American cities and their regions emphasize the highly specific nature in which local élites affect development. Nevertheless all four shared two characteristics: the shift of power from individuals to organizations, and the centralization of power in an organizational network (Walton 1977, 212).

We have found this to be generally valid for our study of the ruling groups of southern Ecuador. The traditional power of the individual landed aristocracy has yielded to that of an organizational influence in which urban family groups satisfy class interests in alliance with each other, with the still powerful Church, with national and foreign capital and with the various levels of government.

Walton's second point is that the local élites must be considered in terms of their class interests, their relationships with other classes, and the ways in which they manage to ensure their interests are implemented through public policy action. This we have also found to be a valuable component in the investigation of the ruling groups in Cuenca. The flexibility demonstrated by these groups in their relationships with business, the church and the state derives partly from their restricted room for expansion in a region with inegalitarian social structures and dependence on external forces. In turn, the possibilities for development in the region have been adversely affected by these same inequalities and dependency ties and by the decision to move the locus of economic expansion to the national centres of development in Quito and Guayaquil.

Accumulation's other face: the ruling class

The dicussion of the ruling groups in the region, and particularly of Cuenca, will centre around their activities as appropriators of the region's surplus, and as investors, speculators and employers making use of that surplus. But their parallel role as consumers and promoters of the consumer ethic is also an important indicator of the social, economic and cultural changes taking place in their society.

A crucial part of the recent expansion of capitalism is connected with the diffusion of new behaviour patterns. The changes in consciousness which reflect international values, lifestyles and attitudes are integral to the process of adapting to modern society, particularly among the dominant groups. In this, cultural persuasion, through a modernized media and especially through advertising, inculcates a growing range of new needs. North American and European films, television police dramas and super exploits of the comics' heroes, with their messages of violence and the lauding of a modern, western way of life, plus American and European pop music and clothing styles, form part of a Euro-American acculturation which has, as one effect, the creation of a widening market for modern factory products.

In Cuenca, the demonstration effects are felt and absorbed, above all, by the ruling bourgeoisie. Euro-American-style homes in the suburbs on the outskirts of the city, imported cars, mechanical household devices (only partially reducing the need for low-paid house servants), detergents and canned and frozen goods, imported, or made locally under licence (and bought in American-style supermarkets or department stores) illustrate the extent of their adaptation to the consumerist lifestyle. This is reinforced by their language, increasingly larded with English words, phrases and references. Much of the above argument is implicit in the rest of the chapter, where we examine the activities of a ruling group who are economically, socially and culturally closer to the foreign values and patterns of behaviour than to those of the majority of the nation's people (how many Ecuadorians who can speak, write or read English are equally fluent in Quechua?).[1]

We are speaking here of a class which collaborates with foreign and national corporate capital to exploit the resources, markets and labour power of its own society. It is a subaltern group whose interests, perforce, are bound up with the dynamic of capitalist growth; as Paul Baran wrote, many years ago, it manages the neo-colony for the advanced capitalist societies (Baran 1957, 205).

So, it comes to associate its well-being and security with borrowed lifestyles, and these imitative élites, according to Amilcar Cabral,

> live both materially and spiritually according to the foreign colonial culture. They seek to identify themselves increasingly with this culture both in their social behaviour and even in their appreciation of its values.
>
> (Cabral 1973, 61)

The conversion of the wealthy into consumers of the commodities, fashions, lifestyles and mores of industrial capitalism has far-reaching effects. As we saw earlier, imitative consumption increases imports, involves costs for research and development, increases the influence of foreign capital in the industrial and commercial sectors, leads to the use of capital-intensive technologies, undermines non-capitalist systems of production and reduces opportunities for employment. The deeper the incorporation into the global system the more thorough the interweaving of these processes of uneven and externally moulded development becomes.

Ecuador has only recently entered into this latest phase of capitalist modernization and most of the impact has been felt in the two principal poles of development, Quito and Guayaquil, where the most influential families and commercial groupings are clustered. Cuenca is a distant third, and its élites have encountered obstacles to their full participation in such processes because of the city's peripheral situation within the national framework.

The regional bourgeoisie of Cuenca is, in effect, an intermediate and mediating group in the national hierarchy of economic power which we show diagrammatically in Figure 7.1. The dominant families of Cuenca are not the final arbiters of the fate of their region, though they wield a significant traditional influence there. But while some of them now range at a national level, within the regional boundaries the group's social and economic dominance is probably less than it was 50 or even 20 years ago. The reasons for this will emerge in the following section.

The structures of social dominance are evident, with the city and its region being ruled by a dozen families; of these the Arízaga, Vega, Malo and Cordero Crespo families are in the first rank, while those of secondary importance include the Moscoso, Borrero, Vintimilla, Vázquez Astudillo, Muñoz, Toral and Tamariz families.

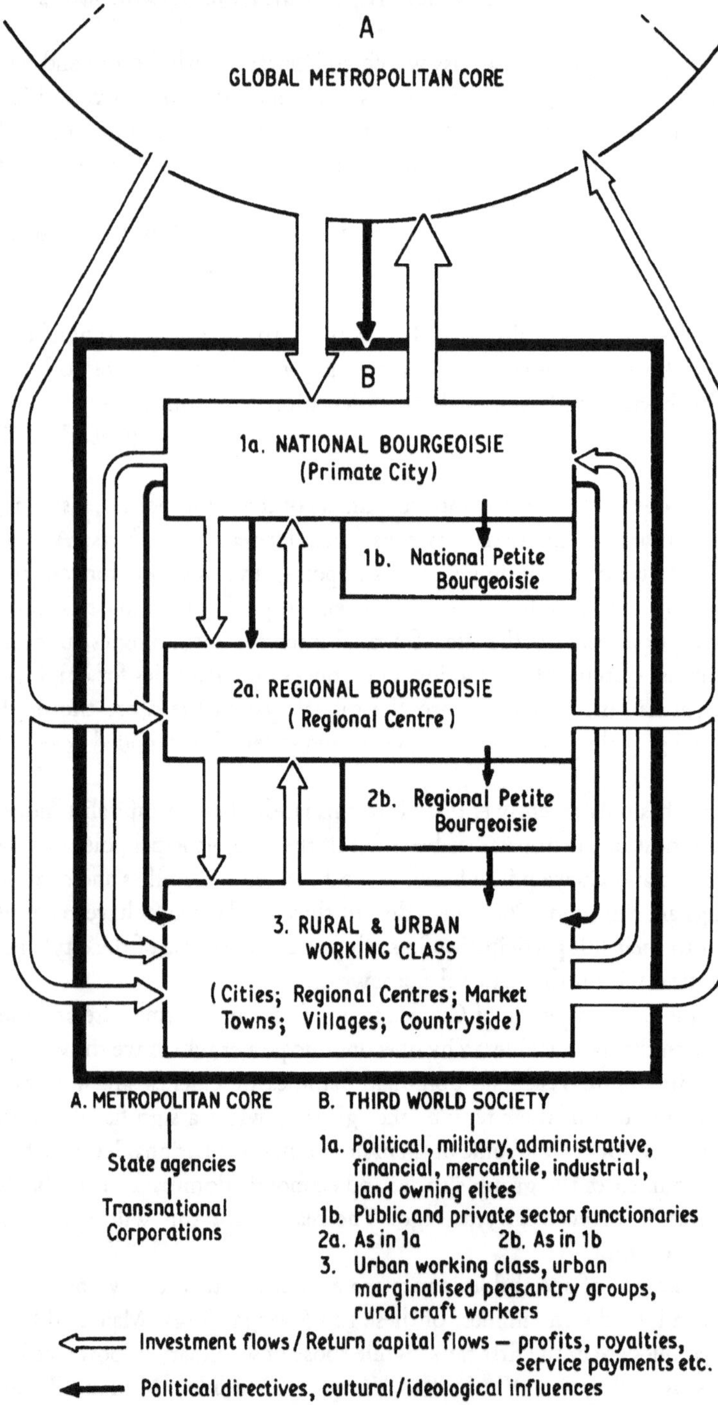

Figure 7.1 Class relations in the urban hierarchy

But the division is now becoming less clear, as intermarriage has led to a certain social mobility. Such marriages at times take on a more than sentimental flavour, as commercial interests are joined by a marital knot. The bonds of class interest are strong among these top families, though some of the younger members now marry outside the network.

The commercial power and influence of the families permeates the three provinces. But they have provided, as well, the important political figures on the local scene and have featured in national politics. Some are also to be found at the head of governmental and other public institutions, in banking and credit agencies, insurance, commerce, transport, building and construction, and in industrial ventures, increasingly linked to foreign capital.

Most do not now derive their wealth or influence from direct ownership of landed estates. They have moved from the land, although rural property is held for leisure activities or its prestige value. They are now an urban ruling group basing their influence on financial and commercial activity, and to a lesser extent, manufacturing. But they still influence agricultural development. They largely control the marketing of agricultural products, extend loans and credit to those working the land, and are the major suppliers of the fertilizer, seeds, implements and other inputs to the direct producers. It is probable, in fact, that their control is greater, and the returns more lucrative, because of their shift from agricultural production with all its attendant risks, to commerce and banking, for they can now more effectively influence land tenure, production and commercialization in the region.

This economic hegemony is reinforced by their vigilance over the ideas and values propagated throughout the society. For this reason the same family names recur in education, and especially in the universities. This is true also in the press where the correspondents often bear the names of privileged families. In all this, economic modernization goes hand in hand with political conservatism.

There is also a strong link between the ruling groups and the Church. Members of the traditional families are to be found in the interstices of the Church hierarchy, from the level of novitiate nuns to that of *Protonotario* of the cathedral, the latter a position of religious, political and social influence. The implications for the extension of ruling class control through the medium of the Church are clearest in the cultural and political spheres, but, as will be seen, there are also economic consequences.

THE ARIZAGA FAMILY GROUP

This general commentary can be illustrated by specific examples. Perhaps the most significant is that of the Arízaga family. A thoroughly urban family group (one must speak of it as such, given its complex kinship and commercial interconnections with other families), it is involved in commercial enterprises tied financially to the Banco del Azuay, also partly owned by the group. This is a convenient arrangement which provides the firms with credit at favourable interest rates, and at the same time allows for effective control over them with limited amounts of investment capital.[2] The family group also has investments in industry, often in collaboration with foreign capital. Inevitably these links have led them onto the national stage, so reducing the importance of Cuenca and the southern region in their total activity.

Politically, their national role is well established; one of the family has run as candidate for the Presidency of the Republic, others have served as ministers in the cabinets of variously hued Presidents. But in a family which has provided Governors of Azuay, leaders of public institutions for the region and senior officers of the region's Conservative Party, the strategy has always been to retain strong social and economic roots, and therefore a firm political base, in Cuenca and the surrounding region.

The Arízaga family group illustrates one fact very clearly: the ability of the ruling class to adapt to changing circumstances. Conservative it may be politically, but its conservatism is no longer that of a landed oligarchy of the Sierra. Its move into banking has reinforced its commercial interests, expanded its rural influence (strengthened further by its strong links to the Church) and has been used to establish financial links with foreign capital (Navarro 1976, 47).[3] And it has transcended the limits of the southern region to gain political and economic power at the national level. The group retains its family links and emphasizes it ties of kinship, especially in the southern provinces.

But, in accepting and adjusting to demands and opportunities of foreign corporations, it has moved beyond mere kinship. It is an economic group responding to the imperatives of mobility and modernization required by foreign capital. The Arízagas form part of the Ecuadorian collaborative bourgeoisie, enmeshed with the transnationals whose penetration in the past decade has greatly expanded the modern sector in Ecuador. They have facilitated this penetration, and are its secondary beneficiaries, supporting the changes wrought by foreign interests while demanding of external capital a part share of the profits.

Such collaboration had, in fact, gone to lengths quite unsuspected until the publication of Philip Agee's *Inside the Company* (1975) demonstrated how far some members had committed themselves to foreign allies to maintain and enlarge their political and economic influence.[4] Although the power of the foreign corporations may often be resented, such ruling groups accept that their future depends to a large degree on maintaining links of mutual self-interest as a counter to any disaffection by their own peasant and working-class population.

The family groups of Cuenca are highly adaptable and have been as flexible as the peasantry in adjusting to a society in flux; they have not only maintained their dominance over peasants, craftsmen and urban workers, but have widened income differences and material inequalities through their capacity to extract surplus in new and more effective ways.

COMMERCIAL INTERCONNECTIONS

A brief summary of their local activity gives some idea of the extent to which they had by the mid-1970s diversified and expanded their activities in the transition from a landed oligarchy to a predominantly urban commercial and financial business class. The three important family groups chosen here to illustrate the evolution come from the dozen mentioned earlier. At the same time, it is not exclusive. Many of the top names do not appear, and some major enterprises are absent. But, as Navarro points out (1976, 76–7), they control an appreciable number of enterprises, mainly in the province of Azuay, and so provide a useful guide to the accumulation process.

The named families – Malo, Vásquez Astudillo, and Cordero Crespo – represent a significant part of the economic power base in the city of Cuenca and its surrounding region. Further, their activities have spilled over provincial and regional boundaries, and their enterprises can now also be found in the Sierra provinces of Pichincha and neighbouring Loja, and on the coast in Los Rios and Guayas. But, outside Azuay, essentially the city of Cuenca, they had among them only one regional enterprise, a bottling plant in Cañar. For the rest, the businessmen in Cuenca either centralized their assets in the city or moved their capital to the two major national centres. The Cordero Crespo group had control of five enterprises and of these, two were in Guayas (Guayaquil) and one in Pichincha (Quito).

The three groups have endeavoured to secure themselves against local competition. By the mid-1970s, each operated over a wide range of activities so as to provide a secure pattern of investment, and the conglomerate structure was complemented by a vertical integration to include agriculture, manufacturing and trading. Moreover, they joined forces to participate in various enterprises including a bottling plant, an agricultural development firm, a mechanic's enterprise, a cinematographic company and the 'Ecuadorian Rubber Company'. And they engaged in joint enterprises with other local family groups, such as the Vega Arízaga family in an importing firm, as well as with large national groupings in Quito and Guayaquil. In the big cities they also acted as partners to foreign capital in a country-mouse/town-mouse relationship. But, at the regional level, their cross links formed 'the existence of economic interrelationships between the three groups, so creating a hegemonic regional economic bloc' (Navarro 1976, 77). What is the nature of this family group capital? Clearly, the economic interests of the ruling groups are basically urban, and each entered into capitalist sector activity individually, in co-operation, or under the aegis of larger national-international corporate groupings. Table 7.1, however, clarifies the structure of the Cuenca family group enterprises, and provides some understanding of their capital investment placements.

The Cuenca enterprises are essentially of a merchant capital nature. Nearly two-thirds of all their capital investments are concerned with relatively secure commercial and hotel activities, whereas less than one-quarter are in the riskier manufacturing sector.[5] Commercial interests are rounded out by minor investments in all the other sectors (except petroleum and mineral extracting), the most notable being in transport and construction.

A number of interesting details in the manufacturing sector also emerge. First, over 90 per cent of the Malo family group's investments is accounted for by subaltern interests in two foreign companies, the Ecuadorian Rubber Company and Industria Llantera Guayaquil, a tyre producer. The Vásquez Astudillo family is less tied, but still has 44 per cent of its manufacturing capital holding a 1 per cent share in the Ecuadorian Rubber Company. The Cordero Crespo family has shown a greater degree of diversity, with its largest single block of manufacturing investment in a plastics company over which it has majority control. Moreover, it ensured a certain measure of security sectorally by diversifying its

Table 7.1 Structure of family-group capital in Cuenca

Sectors[1]	*Malo*		*Vásquez Astudillo*		*Cordero Crespo*	
	Capital (S/.)	*%*	*Capital (S/.)*	*%*	*Capital (S/.)*	*%*
1	—	—	200,000	2.0	50,000	0.5
2	—	—	—	—	—	—
3	2,647,081	24.6	2,311,790	22.6	2,617,870	25.3
4	—	—	—	—	70,000	0.7
5	—	—	—	—	1,000,000	10.0
6	5,906,000	55.0	6,874,000	67.3	6,406,150	61.9
7	1,640,000	15.3	—	—	55,000	0.5
8	9,900	0.1	200,000	2.0	60,000	0.6
9	540,090	5.0	622,000	6.1	99,490	1.0
Total	10,743,071	100.0[2]	10,207,790	100.0[2]	10,358,510	100.0[2]

Notes: [1] Explanation of categories: 1 – agriculture, agricultural services, forestry; 2 – petroleum production, mineral extraction; 3 – manufacturing; 4 – electricity, gas; 5 – construction; 6 – commerce, hotels; 7 – transport and communications; 8 – finance, real estate, publicity; 9 – medical services, film, radio/TV, auto repair.
[2] Percentages rounded.

Sources: Based on information in Navarro (1976), pp. 77–90.

activities. Such vertical integration also strengthened group control over the activities of the agricultural, manufacturing, trade and finance sectors.

The three family groups maintained an influence in agriculture, less by direct means (although the Vásquez Astudillo group had a small investment in livestock production) than by their mutual (minority) interests in various food processing activities. It is also significant that such agricultural interests were concentrated on extensive pastoral livestock activities, although the livestock feedstuffs were based on the production of corn and other grains.

REGIONAL, NATIONAL AND EXTERNAL LINKAGES

Capital accumulation in modern agricultural, industrial and commercial enterprise has been considered so far largely in terms of the southern region. It is not, of course, an autonomous entity, and the point needs to be made more explicitly. First, the process of capital accumulation normally leads to concentration and centralization of economic activity. The location of investment, too, responds to the need for maximizing profitability. By the mid-1970s, foreign corporations in Ecuador had concentrated nearly 90 per cent of their capital in Guayas (45 per cent) and Pichincha (45 per cent), the provinces respectively of Guayaquil and Quito. Azuay received less than 8 per cent.[6]

Cuenca investors still show a concern for their assets in the southern provinces but this may be only a transitional situation. Some Cuenca businessmen have already moved their headquarters to Quito and Guayaquil, in the process transferring the major part of their capital out of the region to the two national centres. Their original head offices then become provincial branch plants, with lessened decision-making capacity and, as intermediaries, channelling capital to the larger centres.[7] This is illustrated in Figure 7.1 which adds a social dimension to the diagrams of these processes presented in Chapter 3.

It might be noted here that theories which argue that economic growth, of itself, will lead to a 'trickle-down' from centre to periphery and a diffusion of benefits from the rich to the poor are, in reality, stood on their head by the process described here. Nor is this an unusual phenomenon; Irma Adelman has come to the conclusion that,

> for the largest part of the development process – corresponding to the transition from the state of development of sub-Saharan African to that of the least developed Latin American countries – the primary impact of economic development is . . . to decrease both the absolute and relative incomes of the poor. Not only is there no automatic trickle-down of the benefits of development – on the contrary, the development process leads typically to a trickle-up in favour of the middle classes and the rich. (Adelman, cited by Goldsmith 1977, 50)

The siphoning of the surplus from the regions and rural areas to the transnationals and national business classes continues. In the present example, the regional growth pole of Cuenca acts as a way-station for such activity; its banks, commerce and industry are intermediaries for the appropriation of surplus and its transfer to the large urban centres.

The removal of head offices from the region to Quito and Guayaquil is one mechanism for centralizing the operations of capital and for increasing regional social disparities. The trend towards ever-greater concentration of capital and control over decision making has its own logic in the system, as Navarro points out when writing of the expansion of one of the national-level enterprises.

> This form of integration, as well as tending towards monopolization at national level, offers multiple advantages, such as the utilization of capabilities (technical, managerial, trading, etc.) acquired in one province, in plants situated in other provinces. (Navarro 1976, 11)

But not even the national 'supergroups' can act independently of the still more powerful foreign corporate capital which articulates the economy around its own growth requirements. The national business groups are even less in evidence in Ecuador than in other Latin American societies so that, except in a few cases, national manufacturers have had to seek some form of collaborative accommodation with United States, European or Japanese companies in joint ventures, or have fitted into niches in the structure built by the transnationals.

National financial institutions, however, have been more prominent, playing an important facilitating role in the new wave of international investment, with banks providing much of the investment capital for corporate expansion. Part comes from foreign loans, but a significant proportion is derived from those processes of surplus appropriation and capital accumulation which are under discussion here.

RECENT CHANGES IN EXTERNAL INFLUENCE

Oil was the catalyst of Ecuador's industrialization. From the early 1970s the country began to follow the path of modern industrial development assisted by direct foreign investment, loans, and aid programmes. From being an export economy dependent upon the locally owned production of a series of cash crops, and subject to foreign mercantile domination of its trade, the country found itself involved in a more complex relationship with direct capital investment.

The earlier dependence upon coffee, cacao and banana exports brought economic instability in its wake. Nevertheless, compared with the new type of investment, mercantile influence and indirect portfolio investment had certain limits in their economic impact. Now, direct investment in the petroleum industry and control over the country's burgeoning modern sectors has widened the control and exacerbated regional, social and sectoral inequalities.[8]

Moreover, the problems of industrialization analysed in Chapter 4 are also observable in Ecuador. Import-substitution manufacturing has opened the way to deeper foreign involvement, while the failures of the state to apply either protective national laws or Andean Pact decrees, mean that outside investors 'have no limitation whatsoever on obtaining credits of national origin, and that, consequently, foreign enterprises finance their activity, in great part, with resources coming from national savings' (Navarro 1976, 41). Yet the very purpose of Decree 24 of the Andean Pact was to ensure that internal savings would be used to strengthen national business and reduce reliance on overseas capital.

Ecuador is thus being modernized, and as this occurs the disparities appear to be growing. Modernization involves restructuring, and the new structures are generally larger-scale, more centralized, and more widespread in their influence. Navarro concludes that:

> both the enterprises controlled by national capital and foreign, have been integrated vertically or horizontally; or they combine both forms . . . so increasing the economic power of the groups under study. One finds . . . both functional and geographic conglomerate growth which ensures the security of investments and the extension of power, and which . . . spills over into other provinces and cities of the country. (Navarro 1976, 20)

The regional implications of the concentration-modernization process at the national scale can now be summarized. As long as Ecuador remained essentially a producer and exporter of cash crops, the impact on the national economy was limited primarily to the commercial influences of unequal trading and pricing practices and to the instability which led to successive booms and slumps in the dependent agro-export economy (Griffin 1969, Chapter 2).

This undoubtedly led, as was noted in Chapter 6, to fluctuations in the export-oriented regional economy of the coast. And the Sierra, supplier of products to the coast and buyer of imports through the port of Guayaquil, felt the secondary effects of such fluctuations. These took the form of reduced purchasing power of the landed oligarchy, increased exploitation of the rural workers and political instability in Quito. But even under such circumstances, production systems and the relations between employer and worker, the rural serfs (or *huasipungueros*) were not radically affected by the activities of commercial capital. In other words, changes in the foreign market situation had a relatively limited impact on social relations of production.

With direct corporate investment the situation is changing. Transnational capital is found in an increasing range of directly productive activities, particularly in the modern manufacturing sector providing consumer goods mainly for the national market. But it is also to be seen in profitable commercial ventures, as well as banking and urban real estate activity, although its participation has decreased in the agricultural sector.

This current phase of capitalist development which Petras designates 'late monopoly capitalism' involves locking-in the profitable sectors of the economy to capitalist production (Petras 1975, 295–6). It is also bringing the peripheral regions into the orbit of foreign capital, dissolving traditional production systems and appropriating surplus for the modern sectors. Because Ecuador has entered this latest stage only recently, the dissolution of traditional, or non-capitalist systems is far from complete, even in the major centres. In fact, in the agricultural sector of the Sierra, and especially of the southern three provinces, non-capitalist production is still alive and active and, for the time being at least, has found ways of coexisting with the modern capitalist sector (see also Poulantzas 1974, 159).[9]

But, if the process of dissolution-conservation is unequal in its incidence, the long-term trend is towards increasing dissolution of non-capitalist systems under the pressure of capitalist competition. In Ecuador, the modern sectors of manufacturing, commerce, mining and urban real estate activities in the two national growth poles are now feeling the principal impact. The corporations are primarily interested in the modern and most profitable sectors of the urban centres of accumulation; labour is most productive here and the surplus for appropriation greatest.

The outlying regions, and the less profitable traditional sectors such as peasant agriculture, however, are also affected by the penetration of foreign corporate capital. Gradual disruption is taking place in the rural society of Azuay and Cañar as those in non-capitalist activity are drawn by seasonal migration into capitalist export agriculture on the coast; into capitalist manufacturing and commerce, as well as into marginal employment activity in the lower circuits of Quito and Guayaquil; and into the more limited range of opportunities in Cuenca. Figure 7.1 has attempted to illustrate the complexity of these movements. The impact of capitalism is not linear, for it does not necessarily lead to the complete dissolution of non-capitalist systems of production while conserving other sectors for the time being; the effects are more complex and subtle.

Finally, the modernization of the society takes a multiplicity of forms, and incorporates many institutions into the process of capital accumulation. In the southern region one of the most influential of these institutions and an active and interested collaborator in the process is the Church.

The Church, surplus and accumulation

> The very lands we all along enjoyed
> They ravished from the people they destroyed . . .
> All the long pretences of descent
> Are shams of right to prop up government.
> 'Tis all invasion, usurpation all . . .
> 'Tis all by fraud and force that we possess
> And length of time can make no crime the less . . .
> Religion's always on the strongest side.[10]

The Church is still the most pervasive, ubiquitous and high-profile organization in the region. Literally and metaphorically at the centre of social life, its network spreads out, incorporating people in the fine mesh of influence which is far more than merely religious. Cuenca's solid, marble-stoned cathedral, whose light blue-tiled cupolas dominate both the central Plaza de Armas and the city as a whole, is clearly the hierarchical focal point of religious activity in the three provinces. But its physical mass is just the most dominating of more than a dozen churches and associated religious institutions whose blue-spired heights break the tiled, horizontal evenness of the colonial city. And, as symbolic evidence of the Church's superiority to the present day, not even the commercial buildings reach the heights of those spires and cupolas. (Indeed, of all Ecuador's cities, only in Quito and Guayaquil is the reverse the case.)

In the broader area of social influence, though, Cuenca's cathedral is really more of a *prima inter pares* than a single, dominating focus of ascendancy throughout the region. It is rivalled by the massive pile of the church which looks down from its hillside site over the small provincial capital of Azogues in Cañar, and also by the startling church on the hill at Biblián, whose soaring spire, seen from the north, forms a compelling point of interest in the mountain valley landscape of central Cañar. The combined urban populations of the two towns is less than 15,000 (less than 30,000 if their peripheral populations are added) yet the social, economic and cultural importance of their churches extends well beyond provincial boundaries.

Every village has its church, invariably the largest and most imposing building in the community, located in the central square. Each reflects in its building materials, architecture and rich, at times exquisite, adornments, either – depending upon one's interpretation – the deep piety of the surrounding population, or the impressive array of means to appropriate the surplus of its parishioners over the years.

A HISTORY OF DOMINANCE

A sense of time is important to an understanding of the church's role and its functioning in society. Nearly 500 years of Spanish and Christian presence lie behind the extensive network of churches which provides the physical evidence of the institution's power.

The evolution of the Church in Ecuador is a central thread in the history of the society since the Spanish took over much of the previous order's religious and moral authority, and converted it to their own purposes. With the private *latifundio* of the *conquistador* and the administrative-military apparatus of the Spanish Crown it was one of the principal foundations of the colonial conquest. From the beginning, religion was 'always on the strongest side'.

Its function was to ensure the incorporation of the conquered peoples into an imposed foreign society, economy and culture. The spiritual thus operated alongside the temporal power, to exercise its legitimizing influence (Stein and Stein 1970, 75–6).

It is scarcely surprising that the Church which accompanied the military conquest, whose intervention was sanctioned by agreements between Rome and the Spanish Crown, and which created vast landed estates while expropriating wealth with legal, indeed divine, right, should have evolved as a conservative institution. The one or two notable exceptions to this help to throw into relief the regressive, yet commercially involved nature of the orthodox Church. Nowhere is this more evident than in Azuay and Cañar where the interpenetration between the Church and the ruling class is so pervasive.

The economic and social consequences of this are evident in the extent of surplus appropriated by the ecclesiastical apparatus in Latin America:

> In general, it is through the good offices of the merchants that the Church realizes, in the mercantile urban nuclei, the surplus extracted from the direct producers – whether thanks to the collection of innumerable ecclesiastical taxes, or to the organization of Indian 'reducciones' and villages.
>
> (Garavaglia 1973, 9)

Today, the Church is a modernizing and successful participant in the evolving commercial and industrial economy of the national bourgeoisie and foreign corporations. This has an important influence on the Church's role in society, for shared material interests create firm and flexible bonds.

As with the ruling class of the region, the Church no longer derives its wealth and power primarily from the direct ownership of landed estates. Nevertheless it continues, from urban and village centres, to claim its share of the fruits of rural activity. The visual evidence of its continuing ability to appropriate surplus can be observed in the constant building and reconstruction which physically manifest the growing gap between the gilded and marble edifices and the adobe houses, lacking water, electricity or drainage, in the villages and surrounding countryside.

Few institutions can have demonstrated as prolonged a capacity as the Church in Azuay and Cañar to extract the region's wealth from its producers. The following examples are a partial and inadequate summary of the Church's ability to manipulate, in Thorstein Veblen's words, its 'vendible imponderables'.

Coros, or religious syndicates,[11] consisting of up to thirty families assume monetary obligations for the maintenance of the church fabric and decorations. In return they ask protection for their lands and crops. During fiestas, *priostes* (festival sponsors) demonstrate their commitment by offering donations or gifts which may cost them a year's savings or the sale of farm animals. At the following fiesta the honour will fall on another *prioste* who will try to improve on his predecessor's performance.

The building of a new church, for example, will entail handing over sufficient land for the building itself. Some parishioners may offer free raw materials, such as a stone quarry for the building materials. The craftsmen and workers in the building will be paid from donations – and perhaps give part of their labour – while the adornments for the interior of the building will be gifts, provided as evidence of the parish's piety.

Religious festivals,[12] and especially those of the month of November are highly munificent. Not only local people but pilgrims from other areas make their donations of cash, and of animals, produce and craftwork for auction. The thousands who go to Azogues in November participate in the processions, make donations and gifts and demonstrate their faith by pinning paper money to the Virgin's clothes. The religious festival acts, then, as a culmination of fund gathering in which faith, fear and social pressure all play a part.

RELIGION AND COMMERCE

How much of the capital thus accumulated is directed to the physical fabric of the Church, to its upkeep, and the propagation of religion? Large amounts undoubtedly are funnelled into such expenditures. But capital is also directed into 'social overhead': charities, hospitals and, above all, education.

The Church's other activities are less apparent although it is generally known that financial support helps to maintain the Franciscan order in Quito, that the Church operates the 'San Agustin' chain of stores throughout Ecuador, and that it has an interest in the internal airline 'SAN'. Navarro, however, indicates the important place of the Church in the national structure of capital accumulation as one of three central groups: 'the foreign [group], concentrator of greatest economic power in Ecuador; the Church, a power constant; and the national bourgeoisie, a modest participant' (Navarro 1976, 2).

Within the region the Church has a small share in the SAN airline based in Cuenca, and a limited holding in the Banco del Azuay, as well as a controlling interest in a wholesale commercial firm in the city (Navarro 1976, 55–6). It has also been steadily moving its locus of investment from the countryside to the cities, and especially into urban real estate. Further, its economic power is continually increasing through its entry into directly productive as well as financial and commercial activities, and it controls part of the communications media – radio and press particularly. This 'allows the religious congregations, besides proselytizing, to mould public opinion towards prevailing social values, in effect towards the economic interests of important members within those very congregations' (Navarro 1976, 55). The Church's participation in banking is significant, so that,

> if the banks are the principal agents in the relation between national and foreign capital, and among those who form 'interest groups', it can be assumed that the religious congregations are not excluded from such economic 'alliances'. Here is yet another reason for the defence of the establishment.
> (Navarro 1976, 56)

Finally, the different sectors of the economy in which the religious groups have investments are developing in a conglomerate pattern, just as in the case of the powerful family groups, and of transnational corporations.

The Church is thus an important component in the process of Ecuador's modernization, and its considerable economic power is a factor in the widening differences between the rural areas and urban growth poles. Because of its collaborative participation in banking, commerce and industry, it is helping to promote a pattern of economic expansion which centralizes activity in the two major cities of the country where growth is most profitable. It is also deeply involved in capital accumulation and the transfer of surplus from the non-capitalist sectors, and so is contributing notably to the 'trickle-up' in favour of the middle class and the rich.

RELIGION AND POLITICAL INFLUENCE

The Church has a political role to play as well, for it remains the most ubiquitous and influential institution in the dissemination of ideas and values throughout the three provinces. The social, cultural and economic changes of recent years have had contradictory effects on different communities in the region. As we have seen, in some cases flexible, innovative responses to the pressures of modernization have been engendered. In others, kinship ties, land scarcity and a general sense of insecurity have encouraged more withdrawn and conservative reactions. Where the latter occurs, the role of the traditional Church has been one of the more important influences in cementing that conservatism.[13]

It is understandable, therefore, that the family groups of Cuenca, lacking a landed oligarchy's direct influence over the peasantry, should co-operate with the Church to try to maintain their social and political influence in the rural areas. An informant in Cuenca explained the *modus operandi* as follows:

> The priests exert a moral influence over the peasantry and lower middle class, and politicians take advantage of this at election time. The candidates keep on the side of the priests, and give money to these religious intermediaries to organize and obtain votes in the elections.[14]

Two observations might be made on these political aspects. The first is to refer to the Steins' point that,

> examination of institutions and practices utilized by an élite and applied to the masses at the local level . . . reveals the operation of a political system far more nakedly than, for instance, an examination of the structure of judicial appeal.
> (Stein and Stein 1970, 76–7)

They cite the examples of debt peonage in India, and *pongaje* or obligatory domestic service by Indians in highland Peru. To these can be added *huasipungo*, Ecuador's Sierra equivalent, and, since the formal abolition of *huasipungo* in the mid-1960s, the continuing informal control by the Church and the urban ruling groups.

The second provides a counterpoint. Although the priests hold great power in the Sierra, the situation is different on the coast where the clergy, according to local informants, appeared to be 'almost without influence'. As the peasants move down to work seasonally or permanently on the banana or sugar plantations, or in the cities, they experience different material conditions and social relationships, and their attitudes gradually change. They become 'less respectful' – or, put another way, 'more awake'. The clergy, it seems, find many of them 'insolent', 'aggressive' and 'rude'. The irony of this is that the Church itself is a principal actor in creating the economic and social conditions in which peasant and artisan must migrate to seek employment in the more prosperous areas of growth. They are, after all, following the same path as the surplus extracted from them.

THE OTHER CHURCH

This section should not end without mentioning the activities of the progressive and popular Church. The discussion has focused up to now on an institution which has evolved over the centuries as a bastion of an often oppressive and conservative social and political order. Yet, within its structure, there are groups of clerics and laymen who, while adhering to their faith, reject the exploitative aspects of the traditional structure.

Educational work, assistance in organizing co-operatives and other forms of community action,[15] 'concientizacion' or raising the levels of peasant awareness, reveal the commitment of these groups of clergy and their lay partners.

The state and regional development

> The State's function
> was to give food to the people
> The earth belonged to those who worked it
> and not the latifundista
> And the Pleiades guarded the cornfields
> there was land for all
> (Ernesto Cardenal 1972)

THE STATE AND RULERS

Poulantzas argues that the state in capitalist society is an expression and crystallization of class power (Poulantzas 1974, 164) so that a crucial relationship in Third World societies is that between the state and local and foreign capital. Poulantzas also distinguishes between national and comprador bourgeoisies. The first, often in competition with foreign capital may, in fact, adopt positions which ally it with the majority in national liberation struggles against foreign corporate capital.

The comprador bourgeoisie, on the other hand, lacking its own base of capital accumulation, operates 'as a simple "intermediary" of foreign imperialist capital . . . and is entirely held in fee by foreign capital' (Poulantzas 1974, 165–6). The possibility of an alliance with the majority of the population is, thus, less likely. Petras, as we also saw earlier, similarly delineates three types of class alliances whose purpose is the accumulation of capital: 'dependent neo-colonialism', 'national developmental' and 'national popular' (Petras 1975, 299–301).

Within the framework suggested by Poulantzas, Ecuador is more representative of a comprador than a national state, while in the area of class alliances for capital accumulation outlined by Petras, Ecuador's 'dependent neo-colonial' regimes have generally opted for a strategy of collaboration with the technology, expertise and capital offered by transnational corporations.

NATIONAL REDISTRIBUTION POLICIES

In 1972, the military seized power. The platform of the new regime was spelled out in a document published just after its takeover; it would be 'Revolutionary; Nationalist and for Autonomous Development; Social Humanist; Disciplined' (Government of Ecuador 1972, 3). A direct attack was levelled at state corruption, inequality, profiteers and the oligarchy. Foreign control, too, was challenged by the government which intended to 'eliminate the country's dependence – economic, political, social, cultural, military and ideological – on the great international decision-making centres' (Government of Ecuador 1972, 4).

In regional development, the proposals aimed at harmonious development, while giving special assistance to 'laggard' regions. Policies would be geared to allow popular participation in the elaboration and execution of decentralized regional planning; to decentralize industries producing for the national market to poles of development outside the primate centres; and to strengthen regional planning within the framework of national planning (Government of Ecuador 1972, 55).

The same philosophy also underlay the more detailed *Integral Plan of Transformation and Development 1973–1977*, drawn up by the National Planning Board. The approach of the new government, it claimed,

> is to consider development as a process not just limited to promoting high rates of economic growth. It also implies a profound structural transformation intended to modify the traditional behaviour of the national economy and society; that is to say, to identify and carry out economic policies promoting local involvement in national development, and offering permanently deprived groups a greater degree of participation in the elaboration and execution of the decisions which affect them. (Junta Nacional 1972, XI)

This last offer would have amounted to a minor revolution in itself; the Planning Board later characterized the 'deprived' or 'marginal' groups in such a way as to include over 50 per cent of the economically active population in the category (Vicuña 1975, 3).

Redistributive policies based on such structural changes would demand 'a better distribution of income and a greater regional balance . . . the need for agrarian reform and the defining of projects throughout the nation's territory' (Junta Nacional 1972, XI). Regional programmes for transport and communications, small-scale industry and craft activities, drinking water and sewage, health and education would all be included. The same approach runs through the specifics of the Plan; urban-rural relations, agrarian reform, industrial decentralization, education, health and a greater degree of popular participation were all part of the strategy to correct the existing and growing social and geographical inequalities in the country.

The difficulties faced in translating this philosophy into effective action demonstrate the divergence between the plans of a liberal bureaucracy and the reality of social and economic power. The regional family groups and national 'supergroups', and foreign capital attracted by the petroleum-based boom, were, in fact, to set the real parameters of Ecuadorean state strategy. It is not surprising therefore that the 'Revolutionary Nationalist' military government of 1972 should have become the conservative regime of 1975. By then it had lost its more radical nationalist members, watered down its earlier philosophy and relegated the proposals for reform to the archives.

The social and regional distribution policies of the government marked a radical expansion of state intervention (or, at least, intended intervention) in societal organization. Nevertheless, they were not a complete departure from traditional activity by the Ecuadorian state. The National Planning Board, founded in the early 1950s, had drawn up four previous plans; each succeeding effort had tried to broaden the sectoral range of its interest and the later plans had incorporated social, regional and agrarian concerns into their prescriptions (Bromley 1977, 82). Equally, as Bromley points out, however, the seriousness of the state's commitment to such planning, was usually in doubt given the failure to implement them effectively, or, in some cases, the outright ignoring of their existence.

Attempts to encourage regional decentralization of economic activity have had a stronger impact. CREA has had some success, with the assistance of central legislation, in attracting industries to the south – particularly, as we shall see, to Cuenca. Others, including the CRM in Manabí, OIPE in Esmeraldas, the extensive CEDEGE based in Guayaquil and PREDESUR, are part of the mosaic of institutions which have grown up in a higgledy-piggledy fashion to try to deal with the pressures from distinct cities and regions in the country.

Doubtless, the expenditures of such institutions, alone or in collaboration with the central government and international organizations, have managed to decentralize some of the benefits of growth. They have also been of value to the central authorities in developing local productive resources, controlling local populations and reducing migration to the major centres. The control objective is of particular significance in Ecuador. Consultation normally takes place only between planners and local élites, for, as Bromley has noted:

> 'Development' is usually viewed as something which is handed down from above, rather than generated from below, reflecting the persistence of paternalism at all levels of Ecuadorian government, and the marginalization of the majority of Ecuador's population from any participation in decision making. (Bromley 1977, 102)

The social and regional redistribution strategies of the military government had to face long-standing obstacles to any real transformation of the planning process. To this can be added the new circumstances of an oil-generated economic boom in the national economy. Growth and modernization, stimulated by petroleum earnings; intensified capital accumulation for urban-centred industrialization; the expatriation of capital in the shape of profits, royalties and interest payments; and the growth of consumer goods imports, were important components of Ecuadorian development. Such a pattern allowed little scope for structural change and programmes of social or geographical redistribution, so that the outcome of regional development projects initiated by the National Planning Board and United Nations development agencies was, at best, inconclusive.[16]

CREA AND REGIONAL GROWTH

Nevertheless, ripples of the passing interest in regional and social development were felt in the southern region, particularly with the joint work of one United Nations regional development team and CREA. This was, however, just another phase in the longer term activity of CREA, which, since its founding in 1958, has proved to be one of the few regional agencies capable of coping with the changing regimes and policies in Quito.

CREA seems to have retained its independence of national politicians and government officials because of the strong support it continued to receive from Cuenca's ruling groups. The Centre's main interests are reflected in its administrative subdivisions – agriculture, infrastructure, livestock, colonization, industry, human development, crafts and co-operative development,[17] but the budget of the 1960s and 1970s tended to concentrate on projects which benefited the economic interests of the ruling groups. More than 45 per cent of the budget in those years went to colonization projects, mainly in the resource frontier of Morona Santiago, and another 27 per cent to infrastructure.

There was a clear disparity between the technical and infrastructural expenditures and social projects. Public financing from Quito aimed to subsidize local development and prepare the ground for the schemes of the principal family groups in Cuenca. Thus, the largest portion of such expenditures was channelled into the area surrounding Cuenca, and the link route between the city and its resource base, the province of Morona Santiago (see CREA 1976, 82). Apart from a small sum paid directly to the Provincial Council of Cañar in the mid-1970s to meet part of the road-building needs, little investment in that province's development was made available (UNOTC 1974, 66).

Nor did the Centre establish effective links with other development institutions and groups at the local level. According to the United Nations development project report:

> This has resulted in a certain breach between the CREA and other local organisms. Such a position is reinforced when the somewhat technical approach of the Centre, which is increasing even more with the dominance of technical programmes in the new budget, is taken into account. (UNOTC 1974, 67)

The contradictions of a development strategy which seemed to avoid dealing with the social and spatial disparities are captured by three commentaries, the first of which appeared in the United Nations report:

> For example, in the case of . . . the Agrarian Reform law, what will be the Centre's principles of action, especially when confronted by the powerful . . . landowners, and equally the demands of the landless or those who possess insufficient land? Such questions will not be avoided or resolved by . . . pragmatism. (UNOTC 1974, 68)

Some of the same concerns about the technical bias and class orientation of the Centre's previous strategies can be detected in the statement of a military director:

> In the first place, planning is a political process; we must construct a new society where there will not be some Ecuadorians preying on others. The construction of bridges and roads is for all men. All have a right to dignity; it is not the property of any one ideology. On the specific tasks of development, CREA should not try to avoid problems but to publicize them openly and frankly.[18]

A columnist, writing in *El Mercurio* of Cuenca, suggested that the Centre should belong to the people, not to a small group offering charity to the peasantry and the poor while pursuing its own interests. CREA should:

> provide the means so that the peasant can discover through material works that there is a superior objective. In making a school, an irrigation channel, a co-operative, agricultural improvements . . . without paternalism, without concealed agendas, without obscure scheming, these are the instruments that will help the peasant to discover himself as a man, collaborating in and determining the road that his class must utilize for its authentic liberation.
> ('Voz del Campesino' 1976, 4)

All three commentaries demonstrate an awareness of the need for conscious change, but they miss an important point: CREA was created primarily to provide for the business interests of Cuenca, rather than the popular needs of the region. It was not intended to act independently of the family groups, or become a neutral arbiter among the social classes and zones of the region. As a facilitator of private sector activities, its prime function was to underwrite the public works required by commercial, financial and industrial investment. The United Nations report observed that:

> From all this, one comes to the conclusion that CREA exists and functions essentially as a promotional organization for the city of Cuenca and its agricultural hinterland in the Province of Azuay. (UNOTC 1974, 66)

But even this is not quite exact, for it does not specify *which* Cuenca CREA serves, or which social and economic interests *within* the city. CREA was not acting on behalf of Cuenca in general but of the city's ruling family groups, and its functions were perceptibly changing in response to the requirements of the more dynamic sections there. It still assisted landowners in northern Cañar but, increasingly, its heavy expenditures went to infrastructural works helping to subsidize urban investors' costs.

In this CREA was responding logically to the needs of capital accumulation: to provide a propitious climate in which to invest, an appropriate infrastructure, and a constant supply of low-cost labour. Its activities were, and are, a necessary complement to the growth and modernization of the region, both in production and consumption, and to the activities of the private actors which we have described in earlier sections.

Finally, to situate these events in the broader context of national development, the experience of the region appears to provide further confirmation of the argument suggested by de Mattos (1982, 79) that even when the state spends appropriated surplus in non-metropolitan areas, this still tends to benefit the ruling groups of the central sub-system. Cuenca, with its modernized environment, has become a more effective transmitter of capital to the centre and, at the same time, is a growing consumer market for the centre's goods. And both these changes are occurring at the expense of local small producers and their families.

8

Conservation-persistence in the two-circuit system of Hong Kong: a case study of hawkers

Introduction

In Chapters 6 and 7 we have shown how the urban institutions and ruling class interact together to extract the surplus from rural regions and funnel it upwards to the nation's capital. Those two chapters focused on the process of accumulation as it occurs within the urban heirarchy and its rural hinterland. In the next two chapters we shift the emphasis to the study of intra-urban relationships that exist between large-scale capitalist production supported by the state and the varied types of petty capitalist production and distribution, as well as the emerging proletariat.

In Chapter 3 we outlined the theory of conservation-dissolution developed by Bettelheim to explain the highly uneven penetration of capitalism, resulting in the persistence of petty capitalism in some cases and its dissolution in others. It will be recalled that we emphasized the considerable advantages to the capitalist and state sectors that accrued from this persistence as well as the significant role that petty capitalists played in preserving their activities. This was further elaborated within the two-circuit system of Third World city economic activity outlined by Santos (1979).

The present chapter looks at a group of petty commodity distributors who have remained remarkably persistent in the face of a dramatic economic growth of the capitalist sector, and shows how this persistence has been of value to the state and the capitalist classes. It is based upon fieldwork carried out between 1969 and 1972. While this information is now dated, the detailed data of the case study remains of great value in testing the relationships between the two circuits.

In Chapter 9 we look at a different segment of the working population which has been rapidly moving from a precapitalist to a capitalist mode of production as factory proletariat. We illustrate these processes by reference to investigation of Malay female workers in Penang, Malaysia based on research currently in progress.

The context

As discussed in Chapter 5 (pp. 93–4) Hong Kong is an anomaly. It is an export-zone economy whose economic growth depends on the role it plays in the international capitalist system. One important point needs to be added to this statement. Hong Kong's proximity and relationship with China greatly affects its internal economy, as the decline of the property market in 1981 dramatically showed. The status of Hong Kong after 1997 now appears to have been resolved by a treaty between the governments of the People's Republic of China and the United Kingdom concluded in 1984. Under this treaty, Hong Kong will continue to remain a 'capitalist enclave' for another 50 years. This should give business in Hong Kong a much-needed boost. Investment, particularly by the Japanese, has already begun to grow.

Despite these developments, there seems little doubt that capitalist expansion in Hong Kong in the last 20 years has been remarkable, and has brought about the dissolution of many petty commodity activities, particularly in agriculture, services and some industries. But other activities have persisted, particularly in retail distribution. Particularly important in this respect are the street vendors with their wide range of commodities. The next section presents a detailed study of these resilient petty capitalists and their relationships with the state and large-scale capitalism; this study is conceptualized as two main circuits of economic activity, an upper circuit of large and middle-sized economic enterprises supported by the state, and a lower circuit made up of small-scale capitalists.

We must stress three caveats against the use of these simple models. The first relates to the dualism implicit in the model of the two circuits, which assumes polar opposites (e.g. capital-intensive/labour-intensive), and does not allow for gradations or a continuum of economic features running from one circuit to the other. The two are inextricably linked and are constantly affecting each other, as for instance when the upper circuit 'captures' the labour from the lower circuit by offering higher wages; or when upper circuit goods replace those produced in the lower circuit. In these cases, lower circuit activities would either be replaced by the upper circuit, begin to assume certain economic features of the upper circuit, or else stubbornly persist in what appears to be an 'irrational' fashion. The data presented later indicate that hawkers, in the lower circuit, have reacted in all three ways to such developments. The reasons for these reactions appear to be the result of a cluster of determining factors, including the type of ecological pattern and commodities sold, and the personal factors. Despite the problems implicit in the use of polarized models, it may still be argued that they do provide concepts to help analyse real situations.

A second argument against the two-circuit model is that it does not take sufficient account of cultural distinctiveness. This argument is particularly important with respect to Chinese behaviour and organization. Some economic anthropologists, concerned primarily with analysing social relationships which structure the systems of production and distribution, have suggested that there is a 'Chinese way' of doing business and organizing economic activity. In a short but stimulating introduction to *Economic Organization in Chinese Society*, Willmont has singled out several elements of the 'Chinese way'. Among them he finds significance in the 'compartmentalized' nature of traditional Chinese enterprises in which production and consumption are carried out by a great many small units, and in the persistence of particularist relations in economic interchange. He says:

> What seems to be uniquely Chinese is the *conscious particularization* of economic relationships, the attempt to develop a multiplicity of ties between individuals associated by economic transactions. Professor Ward has described this as the strong preference for multiple rather than single stranded relationships. To point out the differences by hyperbole, one might say that economic transactions occur along particularist lines in all pre-industrial societies, but the Chinese develop particularist links between those involved in economic transactions. (Willmont 1972, 5)

A great deal of evidence presented by various writers in the book supports Willmont's contention. De Glopper writes of the importance of *hsin-yung* (trust in business) relations in a Taiwanese market town; Silin of the need for a Hong Kong market dealer to have good *kanch'ing* (rapport) in transactions with other market dealers; and Ward of personalized methods of management in a small Hong Kong factory. But it is not clear in these examples what makes such a pattern of behaviour different from that recorded for other societies in which personalized relations govern business relations.

The literature is full of examples of similar patterns: Mintz' discussion of *pratik* in Haitian economic relations (Mintz 1961); Davis' analysis of *suki* relations in the Philippine marketplace (Davis 1968); and many other studies, admirably synthesized by Belshaw (1965). Of course, if it could be shown that such particularist relations persist as a major influencing factor on economic behaviour in the upper circuit among Chinese in Hong Kong, some rethinking of that part of the model would be required. But, for the purpose of this study, Willmont's speculations reinforce the model of social relations in the lower circuit.

A final caveat relates to a major debate that has characterized the field of economic anthropology – namely the question of whether there are different forms of economic conduct in tribal, peasant and industrialized societies which demand different techniques of economic analysis. In a sense, in putting forward a model of the two circuits of the urban economy as a basis for the analysis of hawkers, we are raising some of the questions which have characterized this debate, since it might be argued that economic behaviour of the members of the lower circuit is so different that it demands a different analysis. In simple terms, we ask the question, 'Is the hawker an economic man?'

Alice Dewey, considering the same question with respect to traders in the Javanese context, argues positively on this point, noting that people have to reconcile personal relations with the need to make a profit and that they accomplish this in a perfectly rational manner. (Dewey 1962). We believe the hawkers of Hong Kong behave in the same manner. They have a good idea of profit, and have a considerable capacity to attempt to accumulate capital. Their incomes are affected by a framework of personal, operational and locational constraints that exist in the total economic system; they attempt to overcome these restraints within their available limits. The hawker then has capitalist motives without all the resources of the capitalist.

Hawkers in the Hong Kong–Kowloon metropolitan area: a case study 1969–71

Information on the historical evolution and characteristics of street vendors in the Hong Kong–Kowloon metropolitan area has been very limited, but during the period 1969–71 a variety of data was collected by one of the authors (see McGee 1973) and in other surveys. The ensuing description of the hawkers is based upon these data sources plus surveys conducted by the Hong Kong Government Resettlement Department (1969), Andrew Lu (1972) and the Hong Kong Government Mongkok City District Office (1968).

In mid-1969, at the beginning of this research into hawkers, there were no comprehensive figures on the numbers, distribution, and types of hawkers throughout the urban areas of Hong Kong–Kowloon. The Urban Council estimated that there were some 70,000 of which more than 30,000 were unlicensed.[1] While this number represented only a 'guesstimate', the Urban Services Department had prepared maps indicating the main concentrations of hawkers. Figure 8.1 shows that the main concentrations were located in the older, high-density areas such as Sheung Wan and Yaumati, although they could also be found in the resettlement housing areas such as Shek Kip Mei, and even in the new town areas like Kwun Tong. Within these areas the hawker preferred

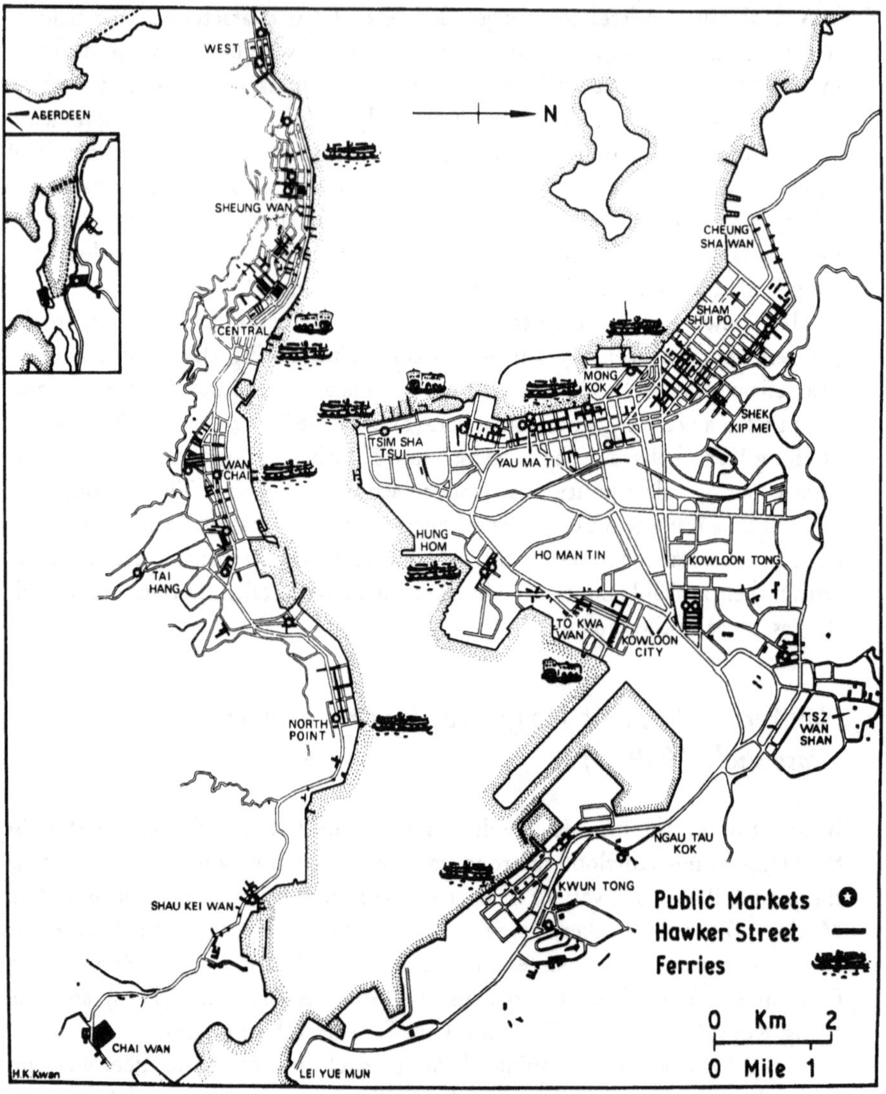

Figure 8.1 Hong Kong–Kowloon metropolitan area: hawker distribution, 1968

locations close to public markets, as shown in the Figure, or in streets of high pedestrian density such as Li Yuen Streets East and West in the Central District. Where possible, they found it convenient to locate close to ferry terminals or other transportation modes which acted as attractions for the population. The areas of high-income housing such as the Peak, Mid-Levels and Kowloon Tong were largely free of hawkers, except for itinerant hawkers selling meat, vegetables

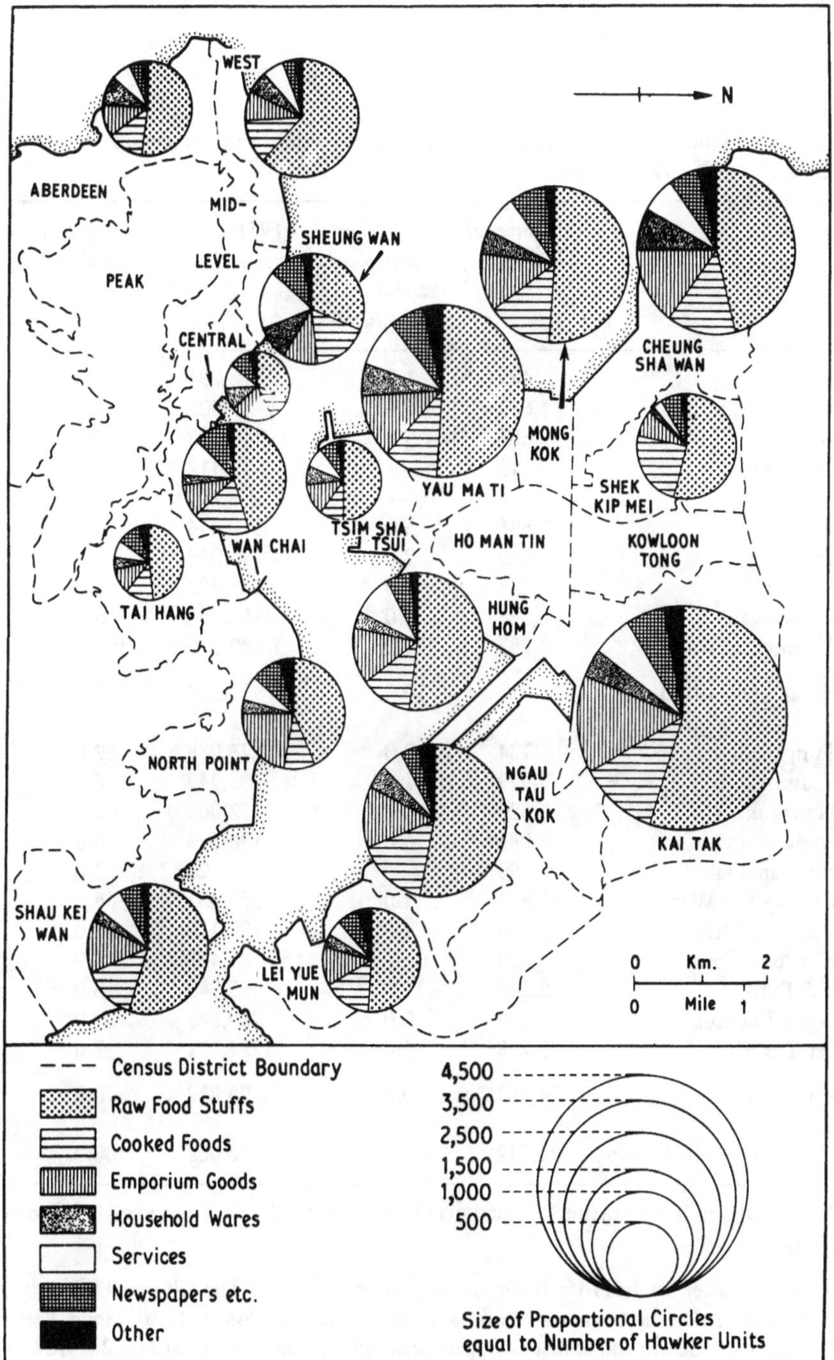

Figure 8.2 Hong Kong–Kowloon metropolitan area: distribution of day-time operative hawker units by commodity sold (census districts, 1971)

Table 8.1 Number and percentage of operative hawking units by census district in Hong Kong–Kowloon, 1971

District	*Number of hawkers*	*% of Hong Kong–Kowloon hawker population*	*1971 population*	*% of total population*
Central	485	1.0	22,794	1.0
Sheung Wan[1]	1,619	5.0	67,907	2.0
Western	1,607	5.0	146,202	5.0
Mid-Levels	12	—	47,112	1.0
Peak	5	—	8,241	—
Wanchai[1]	1,586	5.0	142,844	4.0
Tai Hang	655	2.0	94,040	3.0
North Point	1,438	4.0	176,492	6.0
Shaukeiwan	1,826	5.0	162,456	5.0
Aberdeen	982	3.0	109,116	3.0
Hong Kong Island	10,215	30.0	977,204	30.0
Tsim Sha Tsui	734	2.0	73,988	2.0
Yaumati[1]	3,979	11.0	203,749	6.0
Mongkok[1]	2,820	8.0	172,006	5.0
Hung Hom	2,081	6.0	188,572	6.0
Ho Man Tin	90	—	77,125	2.0
Cheung Sha Wan[1]	3,883	11.0	259,431	8.0
Shek Kip Mei	1,216	4.0	190,138	6.0
Kowloon Tong	30	—	21,403	1.0
Kai Tak	5,846	17.0	555,543	18.0
Ngau Tau Kok	2,590	7.0	230,571	7.0
Lei Yue Mun	1,228	4.0	222,331	7.0
Kowloon	24,497	70.0	2,194,857	70.0
Hong Kong and Kowloon	34,712	100.0	3,172,061	100.0

Note: [1] District's proportion of total hawkers exceeds district's proportion of total population.

Sources: Hawker Study Programme, Social Research Centre, The Chinese University, *Preliminary Tables on Operative Hawking Units, By Census District, 1971*; Hong Kong Government, Census and Statistics Department, 1971, Census Circular No. 2/71, *Hong Kong Census Population By Sex and Tertiary Planning Unit, 1971*.

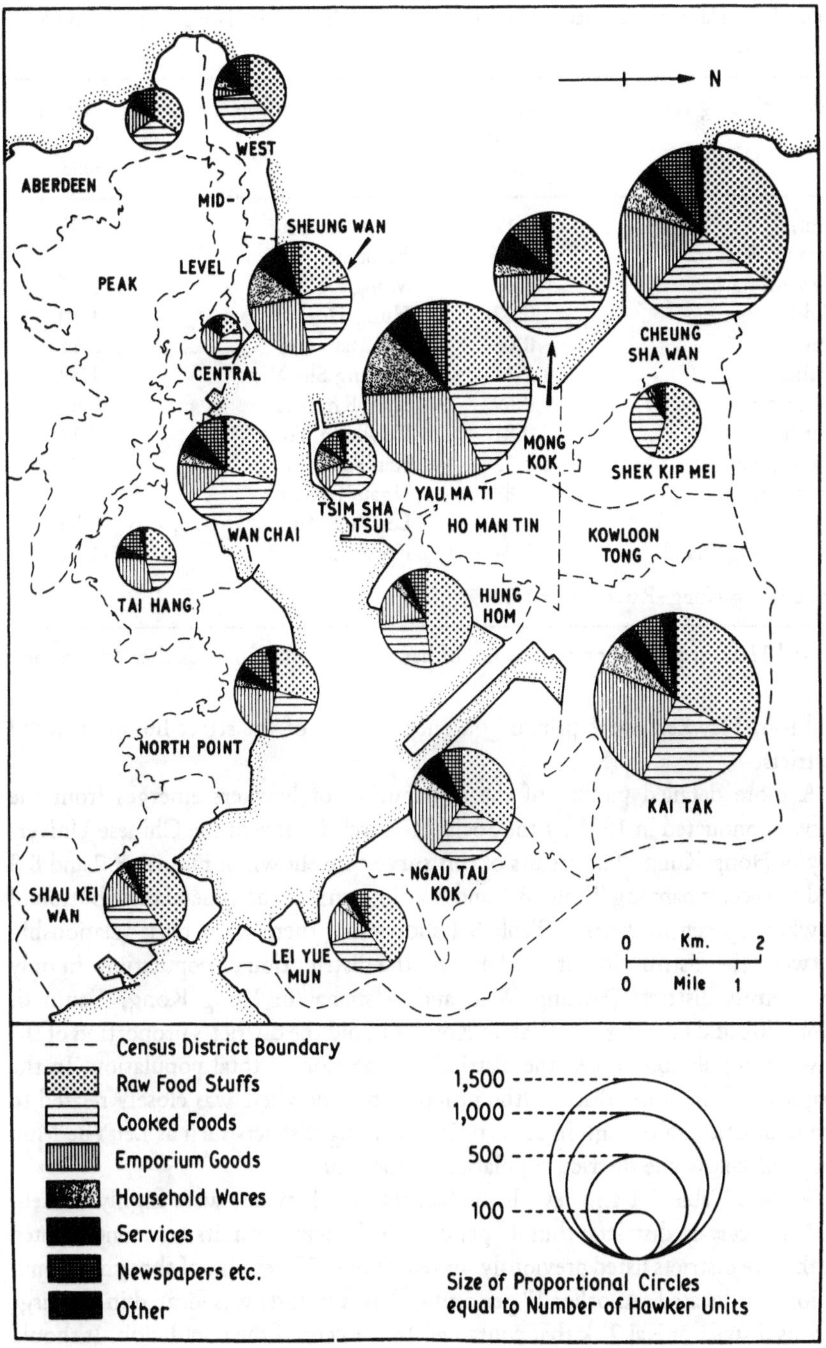

Figure 8.3 Hong Kong–Kowloon metropolitan area: distribution of night-time operative hawker units by commodity sold (census districts, 1971)

Table 8.2 Hawker/population ratios[1] by census district in Hong Kong–Kowloon urban area, 1971

District of Hong Kong	*Hawker/ population ratio*	*District of Kowloon*	*Hawker/ population ratio*
Central	2.12	Tsim Sha Tsui	0.99
Sheung Wan	2.38	Yaumati	1.95
Western	1.09	Mongkok	1.63
Mid-Levels	0.02	Hung Hom	1.10
Peak	0.06	Ho Man Tim	0.11
Wanchai	1.11	Cheung Sha Wan	1.49
Tai Hang	0.69	Shek Kip Mei	0.63
North Point	0.81	Kowloon Tong	0.14
Shaukeiwan	1.12	Kai Tak	1.05
Aberdeen	0.89	Ngau Tau Kok	1.12
		Lei Yue Mun	0.55
Hong Kong Island	1.04	Kowloon	1.11
Total Hong Kong–Kowloon = 1.09			

Note: [1] Hawker population ratio = no. of hawkers per 100 census district population.

and fish. Hawkers were primarily a phenomenon of the street-life in crowded districts.

A more detailed picture of the distribution of hawkers emerges from the survey conducted in 1971 by the Social Research Centre of the Chinese University of Hong Kong. The results of this survey are shown in Figures 8.2 and 8.3 and the accompanying Tables 8.1 and 8.2. In terms of the general distribution of hawkers by census district, Table 8.1 shows that there was a close relationship between the distribution of hawkers and the distribution of population. In only five census districts (Sheung Wan and Wanchai in Hong Kong, Yaumati, Mongkok and Cheung Sha Wan in Kowloon) did the district's proportion of the hawker population exceed the district's proportion of total population. In the majority of the other districts the proportion of hawkers was closely related to the population, although in some of the outlying districts such as Lei Yue Mun they fell below the district population proportion.

Second, Table 8.1 also reveals the fact that the hawkers were highly concentrated in certain districts; thus 40 per cent of the hawker units was concentrated in the five districts listed previously, in which only 25 per cent of the urban population was located. Another 17 per cent of hawker units was located in the large census district of Kai Tak that contained 18 per cent of the population. It should also be noted that the distribution of hawker units between the two urban areas

Table 8.3 Percentage of operative day and night hawking units by commodities or service sold in Hong Kong–Kowloon urban area, 1971

	Uncooked food	*Cooked food*	*Non-food*	*Service*	*Total*
Day-time	51.0	14.0	28.0	7.0	100.0
Night-time	32.0	27.0	35.0	6.0	100.0
Total	46.0	17.0	30.0	7.0	100.0

Source: see Table 8.1.

of Hong Kong and Kowloon related exactly to the proportional distribution of population. So while hawkers are spread throughout all the urban areas (they are insignificant in the high-income, low-density areas of the Peak, Mid-Levels, Ho Man Tin and Kowloon Tong), in certain areas they are highly concentrated.

Despite the concentration of hawkers in these districts, it can be seen from Table 8.2 that only Sheung Wan and Central in Hong Kong, and Yaumati and Mongkok in Kowloon, have a ratio of hawkers per 100 of district population which is well above that of the ratio for the total urban area, 1.09. Among these districts, Central is the exception, for unlike the others, it has a small population living in an area which was occupied largely by commercial space. The hawkers in the Central district are long-established and serve people from all over the Colony. It is significant to note that the 1.09 ratio of hawkers per 100 of Hong Kong–Kowloon total population is much lower than the 2.00 per 100 of population estimated for other Asian cities in an earlier work by one of the authors.[2]

Figures 8.2 and 8.3, along with Table 8.3, show more detailed patterns of hawker location for day-time and night-time operating units and major types of commodity sold in 1971. The day-time distribution of hawker units is important, for it is during the day that major problems of congestion and obstruction occur. This shows a prevalence of raw foodstuff sellers in the day-time in most districts, although it is significant that this feature does not occur in Central, Sheung Wan, Wanchai and North Point, which have significant proportions of cooked food and non-foodstuff sellers. In the Kowloon urban area only Cheung Sha Wan exhibits a similar pattern. There is a sharp difference in the numbers of hawkers and the distribution of commodities sold by night-time hawker units for here cooked food and emporium goods (principally textiles) dominate.

The two districts of Sheung Wan and Yaumati are major areas of hawker concentration, for there the two largest night markets are located, the 'Poor Man's Night Club', held in the car park outside the Macau Ferry in Sheung Wan, and the Temple Street night market held in Yaumati. Thus, a considerable diurnal variation in the numbers, location and commodities sold in the Hong Kong urban areas imposes a distinct spatial patterning on hawking activity and customer response. Table 8.3 shows that 51 per cent of the day-time hawker units sold uncooked foodstuffs, compared to 32 per cent of the night-time units. Cooked food and non-food sales dominate most of the night-time activity.

While this pattern of diurnal variation in the commodities sold is significant in describing hawking activity in the urban areas, there are also spatial variations in the types of hawkers and the commodities sold which can be shown by a spatial model of hawker activity. Table 8.4, showing the density of hawker and population per acre in 1971, is the final piece of information needed to indicate a model of the ecological *milieu* in which hawkers operate.

Together with the other material presented, it is possible to distinguish five main ecological *milieux*. In Sheung Wan, Western and Wanchai in Hong Kong, and Yaumati, Mongkok, Hung Hom and Cheung Sha Wan in Kowloon, the number of hawkers per acre is well in excess of the figure for urban areas as a whole. Then, there are the low density districts such as Kowloon Tong where hawkers are not significant. In the two districts of Central and Tsim Sha Tsui the hawker density is low, although hawker-population ratios are quite high. Finally, there are the remaining outer districts of the urban areas where the density of hawkers is well below the urban figures. Areas such as North Point with a middle-class population can be distinguished from the resettlement areas with low-income populations which dominate many of the Kowloon districts such as Kai Tak.

Such an ecological division follows closely that put forward by C. P. Lo on the basis of his analysis of the 1966 By-Census data, in which he drew a spatial model of Hong Kong based on the socio-economic data available in the census (see Lo 1972). It is clear from the earlier tables that of all the ecological *milieux* the hawkers favour most the crowded tenement areas with high population densities, which have the poorest housing standards, and the resettlement estates (see Drakakis-Smith 1970; Pryor 1971). A rough estimate would place the number of hawker units in these two areas at approximately 80 per cent of the hawker population in the Hong Kong–Kowloon metropolitan area, once again emphasizing the fact that the hawkers are primarily concentrated in these two crowded environments.

Hawkers tend to conform to one of four patterns within the streets which form their trading space. First, there are *focused agglomerations* such as the hawker concentrations around public markets in the densely crowded tenement areas. Typically such agglomerations cluster about the entrance to the market, but sometimes government control has forced them on to one side of the market in a more regular fashion. Then there are *bazaar-type agglomerations* which cover an area uninterrupted by a central fixed focus like a public market. Such bazaars are often found illegally occupying public lands in resettlement estates. A more

Table 8.4 Hawker unit and population density per acre in Hong Kong and Kowloon urban area, 1971

	Area (acres)	*1971 population*	*Density per acre*	*No. of hawkers*	*Hawker density per acre*
Hong Kong urban area[1]					
Central	259.10	22,794	88.0	485	1.87
Sheung Wan	170.28	67,907	399.0	1,619	9.52
West	376.65	146,202	389.0	1,607	4.26
Wanchai	217.40	142,844	658.0	1,586	7.30
Tai Hang	875.47	94,040	107.4	655	1.33
North Point	997.14	176,492	177.0	1,438	1.44
Shaukiwan	1,189.80	162,456	139.0	1,826	1.53
Total	4,085.84	812,735	199.0	9,216	2.25
Kowloon urban area					
Tsim Sha Tsui	446.62	73,988	165.5	734	1.64
Yaumati	391.38	203,749	521.0	3,979	10.17
Mongkok	274.13	172,006	627.7	2,820	10.29
Hung Hom	499.70	188,572	377.1	2,081	4.16
Ho Man Tin	601.20	77,125	128.3	90	0.14
Shek Kip Mei	846.12	190,138	224.7	1,216	1.43
Kowloon Tong	617.80	21,403	34.6	30	0.04
Kai Tak	3,189.59	555,543	174.1	5,846	1.83
Ngau Tau Kok	1,398.59	230,571	164.8	2,590	1.85
Lei Yue Mun[2]	1,832.98	222,331	121.2	1,228	0.67
Total	11,267.27	2,194,857	194.8	24,497	2.17

Notes: [1] Aberdeen, Mid-Levels, Pokfulam and Peak included; [2] Lei Yue Mun excludes Tertiary Unit 833.

Source: see Table 8.1.

regularized form has been adopted for the off-street bazaars for relocated hawkers.

Thirdly, there are *streets of specialized commodities or service hawkers* who may sell textiles, books or even engine parts. Their location is often related to existing land use, as is the case with the many specialized street concentrations in Mongkok and Sham Shui Po. Finally, there are the *night bazaars* to which we have already referred. Each of these ecological concentrations poses different problems in terms of policy and relations with the state and more developed capitalist sectors.

Hawkers' participation in petty commodity retailing

ECONOMIC ASPECTS

This section explores the question of the hawkers' participation in the lower circuit of the urban economy in three sections: first, the economic aspects of the hawkers' operation, the flow of commodities and their methods of doing business; second, the personal features of the hawkers and their social relationships with other hawkers and the public; finally, a conclusion which discusses hawking as a livelihood and considers its future.

The information generally is analysed according to the two main commodity groups sold – food (principally vegetables and fruit) and non-food (principally textiles and clothing) – each of which, at least in the context of this study, tended to cluster ecologically and to develop different customer fields.

In this section we describe some of the major economic features of the units of operation in the lower circuit. These include the family organization of the unit, small capital and stock, the extensive use of personal credit, close relationships with customers and a lack of fixed prices. The hawker units of operation surveyed in the study did not always conform to this model, in part, it may be argued, because there were factors such as government interference which affected their behaviour.

The form of ownership and labour utilized in the operation of the business was clearly a critical determinant in placing the hawkers within the lower circuit. The survey indicates that 'self-ownership' made up almost two-thirds of the hawker units surveyed. Family ownership represented only some 17 per cent, while partnership, renting and other forms of ownership formed the remainder.

On the face of it, this suggests that the majority of hawker units of operation are not owned on a family basis, but in fact 'self-ownership' generally means that the family operates the unit. Interviews with the hawkers indicated that the response to the question was considerably affected by the fact that the majority of hawkers (94 per cent) had a licence issued by the urban authorities in their own names, and they interpreted this as their personal right of ownership. Thus if the first two categories of ownership – personal and family – were taken to be synonymous with the family basis of the economic operation which dominates the lower circuit, then it can be seen that 80 per cent of the hawker units of operation were organized on this basis. The remaining 20 per cent were significant, of course, but unfortunately detailed data on the form of ownership are lacking, so it is difficult to infer that they were organized on principles closer to that of the modern firm.

A second set of variables relates to the number of people involved in the unit of operation, including the hawker operative. Data collected in the Hong Kong

1969 survey shows that the majority of hawker units in the operation used other labour in addition to that of the hawker himself, and that there was little difference according to the category of commodity sold. This appears to conflict with the results of the Chinese University Survey which found that only 14 per cent of the hawkers used other labour. It is assumed that this sharp variation was the result of different interpretations of the question (see Lu 1972, Table A18).

A third set of economic variables relates to the value of stock, daily turnover, and incomes of the hawker unit. Predictably this is the information which shows the greatest likelihood of error. The interviewers reported that the hawkers tended to underestimate the value of their stock and their daily incomes, but cross-checking with other surveys indicates that the order of the data recorded was broadly comparable. The stock value question did offer substantial possibilities for underestimates for it asked the hawker to estimate the value of his present stock, rather than the daily average value of his stock. This meant that in some cases, if hawkers were interviewed after they had sold a considerable portion of their day's goods, the value of stock would be much lower. Generally, however, the value of stock held by the hawkers at any time was not great, compared to that of shops and department stores. The value of stock possessed by the hawkers was obviously significant with respect to the type of hawker installation.

Comparison of the figures for Hong Kong hawkers in Table 8.5 with stock values recorded for hawkers in resettlement estates and the Mongkok area shows that the Hong Kong hawkers carried stock of much greater value than those in the other two areas, which is supported by the evidence on advantageous locations of Hong Kong hawkers in the major high-order retailing areas.

The data on hawkers income are subject to the same questions of reliability as the data on value of stock. Reporting on the results of a survey of daily turnover and gross profit among a wide variety of stallholders in Hong Kong, Tse commented that some hawkers were unwilling to give true incomes, for they felt disadvantages might follow. This was certainly the case with illegal hawkers and hawkers with high profit margins. On the other hand, those hawkers with narrow profit margins were more willing to give reliable information. Perhaps this is why the data he presented surprisingly indicates very little difference between vegetable and textile hawkers in the amounts of daily turnover ($HK100–200) and in profit margins (10–20 per cent). Clearly the textile hawkers have a greater reason to conceal their real profit, particularly as, unlike the vegetable hawkers, they are not subject to government control of wholesale prices (see Tse 1971).

Whatever the reason, the Hong Kong Survey data indicate that there is a great difference in the daily profit earned among the major categories of commodity sellers, which conflicts with Tse's evidence. Thus, the majority of raw foodstuff

Table 8.5 Hong Kong hawker units of operation by value of present stock and hawker type, Hong Kong, Mongkok and resettlement areas

					Total	
	No.	%	No.	%	No.	%
	Below $HK150		*Above $HK150*			
Hong Kong[1]						
Fixed-pitch stall	125	34	242	66	367	100
Fixed-pitch pedlar	80	54	68	46	148	100
Wall stall	2	3	60	97	62	100
Total	207	36	370	64	577	100
	Below $HK140		*Above $HK140*			
Mongkok[2]						
Fixed-pitch stall	433	43	580	57	1013	100
Fixed-pitch pedlar	439	66	228	34	667	100
Wall stall	11	69	5	31	16	100
Total	883	52	813	48	1696	100
	Below $HK140		*Above $HK140*			
Resettlement[3]						
Fixed-pitch stall	459	36	807	64	1266	100
Fixed-pitch pedlar	573	80	139	20	712	100
Wall stall	159	57	118	43	277	100
Total	1191	53	1064	47	2255	100

Note: To facilitate comparison, certain categories have been combined.

Sources: [1] Hong Kong Hawker Survey, 1969; [2] Mongkok Hawker Survey, 1968; [3] Resettlement Survey, 1969.

hawkers earned a profit between $HK5 and $HK15 a day, a monthly profit in the vicinity of between $HK150 and $HK450, whereas the majority of textile sellers earned between $HK10 and $HK30 a day, giving a monthly profit of between $HK300, and $HK900. The 'other commodities' and processed food categories seem to have earned incomes in between these levels.

Of course, the amount hawkers earned was greatly affected by the number of days they worked each month and the prevailing weather. Thus, rainy weather could cause a considerable fall in their monthly incomes, but, on the other hand, seasonal festivals, such as Chinese New Year, often resulted in a considerable increase in income. On the whole, these low daily incomes forced the hawkers to work long hours, and they were 'open' virtually seven days a week. Indeed almost a quarter worked for more than twelve hours a day, although the availability of family labour to staff the stall during periods of absence of the hawker operative was a major factor in this pattern.

It would seem, then, that the hawkers generally worked long hours with fairly low hourly returns for their input. Comparison with other income patterns in Hong Kong is fraught with difficulties, in part because of different methods of data collection, and in part because it must be assumed that hawker incomes were only part of the total household income, and we have no evidence on the total household income of hawkers. Hopkins' report on a 1968 socio-economic survey of households in different parts of Hong Kong indicates that the average monthly income of heads of household ranged from $HK332 in resettlement estates to $HK462 in private tenement areas (see Hopkins 1969, 5). Certainly then, the incomes of most of the hawkers, with the exception of some of the vegetable hawkers, would appear to compare favourably with these figures, and in the case of the textile hawkers to be more than favourable.

This is borne out to some extent by the response to a question in the Chinese University Survey which indicates that only 50 per cent of the raw foodstuff hawkers felt that the money earned from the stall was sufficient for the family, compared to some 70 per cent of the hawkers selling emporium goods, principally clothing (see Lu 1972). However, the indications are that there has been a general increase in the level of household incomes in Hong Kong since 1968, as well as wages in the upper circuit, and it may be that the hawkers in the food sector have not experienced the same level of increase.

The goods sold came from a wide diversity of sources, but again there was a major difference between the food and non-food hawkers in the flow of goods. The vegetable and fruit hawkers generally got their produce from the Kennedy Town wholesale market administered by the Urban Services Department and, in a few cases, from the wholesale market in Kowloon operated by the Vegetable Marketing Organization. On the other hand, the non-food sellers acquired most of their goods from either factory or wholesaler, or both, and were thus reliant upon upper circuit production. Thus these two commodity groups have different marketing and sales demands which affect the patterns of their purchasing, frequency of delivery and, to some extent, their relationships with individuals in the marketing chain.

The importance of vegetables in the Cantonese diet has often been commented upon, since they form a major part of the household budget of the Hong Kong population. Wong reports that, according to the Household Expenditure Survey, 1963/4, 'the average family in Hong Kong spends 8 per cent of its household food budget on vegetables as opposed to 9 per cent on pork and 13 per cent on rice (Wong 1970, 82). It is customary for each household to consume fresh vegetables at least once a day. The Cantonese distinguish between the highly perishable green vegetables *chap choi* (wet vegetables) and the less perishable vegetables such as cabbages, ginger and potatoes, *kon choi* (dry vegetables).

Vegetable hawkers sell both types of vegetables, sometimes specializing in one or the other, but both demand fairly frequent delivery and neither has any great capacity for storage, although the dry vegetables can be kept for longer periods. Once the vegetables are in the markets, whether they are imported from overseas (principally China) or locally produced (i.e. in the New Territories), they are generally sold by *laans*, firms which receive produce in bulk from certain suppliers to be sold to smaller wholesalers called *chakkas*. The *chakkas* break down the produce into much smaller units, often baskets of 150 catties (90 kg) or less, which are then sold, sometimes through even more intermediaries, to various outlets, principal among which are the hawkers. Silin (1965, 1972) records that there are many variations in this distribution but in all of them there are systems of credit relations which depend upon close personal understandings and confidence. Fruit hawkers get most of their produce directly from *laans* in various parts of the Colony.

With the other major category of commodities, largely clothing and textiles but including a diverse range of household commodities, it is more difficult to establish exactly the sources and pattern of the distribution of goods. The majority of these goods are manufactured in Hong Kong. Some informants in the Urban Services Department were of the opinion that hawkers often received these goods from factories when overruns of items being produced for export occurred. Others maintained that the hawkers were acting as agents selling on commission for the factories, or even as direct selling outlets for factories. At any rate, it did seem that these hawkers received their goods in much the same way as conventional stores, although the systems of credit extension were more personal. They thus played an important role in distributing products of the formal sector, and certainly benefited Hong Kong factories.

THE HAWKER BUSINESS

The success of the hawker business, as with other retailing activities, rests upon the capacity of the hawker to attract customers to purchase his goods or services. This involves the hawker in a variety of decisions concerning the location of his enterprise, the goods he wishes to purchase and sell, the manner of display, the prices to be asked, and the limits of his overhead payments to keep his business functioning efficiently. Almost none of these decisions can be made in isolation from the influences of government control, prevailing patterns of market behaviour and economic and land use changes within the city, so that, in most cases, the hawker operates his business within a framework of constraints which he is powerless to control. Within this framework, however, he can make a number of decisions concerning his enterprise.

Most hawker units of operation had been located in their present pitch for many years.[3] The hawkers chose their sites for a variety of voluntary and involuntary reasons. In the case of the Hong Kong survey, reasons listed as 'other', including 'close to home', 'inheritance' and 'close to friends', formed the major group, while the second largest group came under the heading of 'government direction', including 'relocation' and 'licensed to operate on this spot'. Strictly economic reasons, such as 'large number of customers' or 'good for business' made up the final category. When this data was analysed by type of hawker unit, the fixed-pitch pedlars showed a slightly higher percentage of economic reasons than the fixed-pitch stalls, but this is because of the fact that the stalls have been subject to much more government control. Government influence on the locational patterns of hawkers in Hong Kong has been considerable and therefore this proportion is understandable, but social reasons were still important.

Asking a similar question of a sample of hawkers in Hong Kong and Kowloon, Lu found a smaller proportion of 'government reasons' and a higher proportion of 'economic reasons', but the largest category of reasons given was 'other', in which 'closeness to home' was the single largest category. It is clear that the hawker attempts to cut some of his overhead costs by locating his business as close to his home as possible. This is supported by Lu's additional information that some 75 per cent of Hong Kong–Kowloon hawkers lived within 15 minutes' travelling distance of the stall. The hawker, like the citizen of the pre-industrial city, seeks to bring his home as close to his place of work as possible.

The assertion that the low price of goods is the main attraction of hawkers is seen to be debatable when the customers' reasons for shopping from the hawkers are analysed. In Table 8.6 the responses to this question have been grouped into four main categories in order to distinguish differences in response between food and non-food customers. Under the 'economic' heading come such responses as 'low price', 'better quality of goods' and 'wide variety of choice'. In the 'locational' category, the two principal responses were 'close to home or office' and 'easy access'. 'Social' reasons were primarily 'familiarity with hawker', and the 'miscellaneous' reasons included such responses as 'just passing by', etc. What is important here is the fact that the hawkers are perceived as 'convenience vendors' and as such probably play an important role in cutting down costs like those of distribution, thus helping to restrain costs of living. This was clearly beneficial to the extent that it kept wages down, since this would enable Hong Kong manufacturers to compete more effectively on international markets.

Table 8.6 also shows an interesting contrast between the two commodity purchasing groups, with locational factors influencing the food purchasers and economic factors the non-food purchasers. We have already noted the locational

Table 8.6 Percentage of Hong Kong hawkers' customers by major type of goods purchased and reason for purchase

Reasons	*Food*		*Non-food*	
	No.	*%*	*No.*	*%*
Economic	392	30	469	68
Locational	576	44	91	13
Social	70	5	21	3
Miscellaneous	203	16	94	14
No information	59	4	16	2
Total	1300	99	691	100

importance of the neighbourhood food clusters, which is strongly supported by this table. It seems vitally important for the food hawkers to locate themselves advantageously within the relatively small area from which they draw their customers. The non-food hawkers, on the other hand, can attract customers by offering low prices as a major attraction, although they, too, prefer a location with high pedestrian densities.

Most customers of the Hong Kong hawkers purchased similar goods from other hawkers. From the food hawkers, they bought in such small quantities that it was often advantageous to purchase several different items from different hawkers. The non-food customers also frequently bought from other hawkers, often in different locations. Since low cost was an important factor, they were willing to 'shop about' for the lowest prices. All hawker customers also bought from shops, which indicates the symbiotic relationship that exists between the hawkers and the surrounding stores or market.

There are occasional items in the newspapers which suggest that the owners do find the concentration of hawkers in their vicinity a nuisance. For instance, Ho quotes a letter dated 21 March 1972 to the *Sing Tao Man Pao*:

> We are the shopkeepers of shops along Sao Yeung Choi Street from Argyle Street to Soy Street. Recently every night, there have been many hawkers crowding at the above-mentioned locations, selling things from their carts which block all the entrances of nearby shops. They make the street filled with rubbish, causing unhygienic environments. They also cause obstruction to the pedestrians who cannot walk easily on the pavements. All the shops are seriously affected, especially during the last Chinese New Year.
>
> (Ho 1972, 222)

Lu found that the hawkers questioned in his survey did not believe that the shops cared very much about them (see Lu 1972, 76). In the case of the hawkers in the Hong Kong survey, it seems that they had either established a fairly good relationship with the surrounding shops, or had usurped the possibility of the growth of competing shops.

PERSONAL CHARACTERISTICS

So far we have shown the main features of the economic organization of the hawker unit of operation, the flow of goods to hawkers and then customers, but the features of the operation are greatly affected by the personal characteristics of the hawkers, their reasons for entering the profession and their social relationships with each other, their families, the public and municipal authorities.

Many studies of the personal features of small traders in various societies have emphasized the dominance of women in this occupation, particularly in West Africa (Marshall 1964), the non-Hispanic Caribbean (Katzin 1959), and some parts of mainland Latin America (Tax 1953). Mintz (1971), in a provocative review of the literature, has attempted to show how the family structure and social organization of these societies is often conducive to female participation in these activities, even in situations where the structure of employment is rapidly changing to a dominance of wage labour, as has certainly been the case in Hong Kong in the period since the Second World War.

There is, however, rather conflicting evidence on the actual participation of females and males in hawking occupations. Table 8.7 shows that the 1969 survey of Hong Kong hawkers revealed a dominance of female operatives with a ratio of 96 male operatives to every 100 female operatives. No other survey reveals similar patterns; in fact, the others all indicate a male dominance. It seems highly probable, then, that the 1969 survey of Hong Kong hawkers underestimated the

Table 8.7 Hawker operatives: ratio of males to 100 females

	Ratio
Hong Kong–Kowloon Metropolitan Area 1971[1]	123
Hong Kong Hawkers 1969[2]	96
Resettlement Hawkers 1969[3]	145
Mongkok Day Hawkers 1968[4]	170
Mongkok Night Hawkers 1968	204

Notes: [1] Lu 1972; [2] McGee 1970; [3] Hong Kong Government Resettlement Department 1969; [4] Hong Kong Government Mongkok City District Office 1968.

proportion of male operatives, and this factor should be taken into account in the subsequent discussion of occupational mobility and social relationships.

It is possible to suggest at least three reasons for this underestimation. First, the survey simply recorded data on the personal features of the operative who was looking after the unit at the time of the interview. No attempt was made to get personal details on those not present, for it was felt that these could not be checked and would be subject to error. Because many of the interviews were, from necessity, carried out at times of slack business, it may have been that wives were minding the unit while their husbands were eating or away. A second reason relates to the nature of the survey and the *milieu* in which it was carried out. All the other surveys had been conducted with the knowledge that they were in some way connected with government, and it seems likely that in those cases the hawkers made a greater effort to ensure that the head of the operation – often the male household head – was present at the interview.

A third factor is related to the various types of commodity sellers in the sample for, as we shall show, there was a pattern of sex differentiation in the selling of different commodities. In fact, apart from the night hawker survey in Mongkok, where cooked-food stalls which employ large numbers of males were significant, there was no great variation in these patterns.

On the other hand, if the overall evidence that the majority of Hong Kong hawker operatives are male is correct, then we have an interesting variation on the patterns which prevail in the trading sector of many other societies, indicating the persistence of historical patterns of male hawking in the Chinese-dominated cities, not only in Hong Kong but in Singapore as well. The pattern of sex participation in the broad categories of commodities does, however, indicate that females dominate in the distribution of raw foodstuffs, principally vegetables and fruits. This is decidedly the case among the Hong Kong hawkers where, as we show, these females represent a long-established community. In all other commodity sectors, males dominate in both the prepared food and non-food sectors. Thus the Hong Kong situation does have some parallels with the patterns of other societies, for the prevalence of females in the foodstuff distribution sector is often cited.

The age structures of the Hong Kong hawkers are shown in Figure 8.4, which indicates that while there were some variations among the different groups of hawkers surveyed in selected areas of the Hong Kong–Kowloon metropolitan area, the overall age structure was heavily weighted in the over 40 years age bracket. The resettlement hawkers and Monkok hawkers had the highest proportion of hawkers in the older age bracket, above 50 years, while the Hong Kong hawkers exhibited a marked concentration in the 40–9 years age group. There is little evidence from this data to support the rather common view expressed by some pressure groups that the hawkers recruit a young population

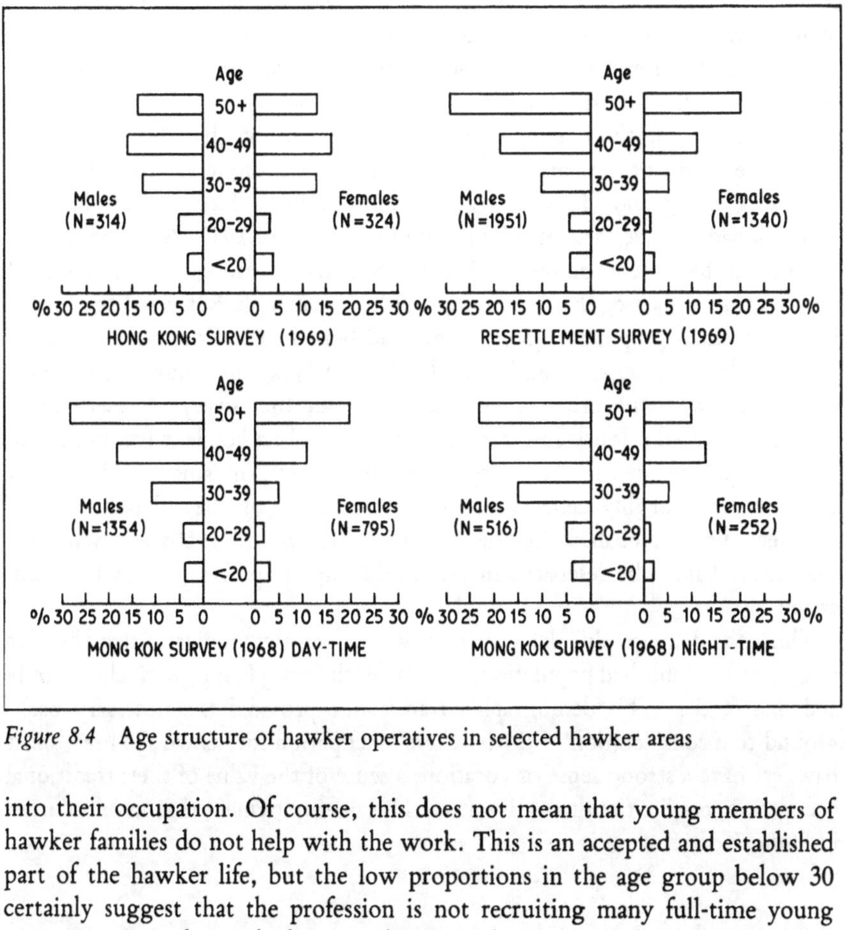

Figure 8.4 Age structure of hawker operatives in selected hawker areas

into their occupation. Of course, this does not mean that young members of hawker families do not help with the work. This is an accepted and established part of the hawker life, but the low proportions in the age group below 30 certainly suggest that the profession is not recruiting many full-time young operatives. Lu's data, which is not shown in the Figure, indicates that only an estimated 11 per cent of the hawking population were below 30 years of age, giving additional colloboration (see Lu 1972, 42).

The survey of Hong Kong hawkers did not record data on marital status or the number of family members, but the information available from the sample of hawkers in the Hong Kong–Kowloon metropolitan area indicates that a high proportion (83 per cent) were married and that, hardly surprisingly in view of the overall age structure, just over 66 per cent had four or more persons living with them. Demographically, the picture of hawkers is of an ageing occupational community, married, with well-established families. This demographic data is supported by information on place of birth and years of residence in Hong Kong. The majority of the Hong Kong hawkers had been born outside the Colony, most of them in the region of Canton and Macau, a smaller proportion in the

Sze Yap region to the west, and in the Chiu Chau areas east of the Cantonese region. Only some 18 per cent had been born in the Colony, a figure almost identical to the proportion of the total population over 20 years who had been born in the Colony at the time of the 1966 By-Census.

Data on the length of residence in Hong Kong shows that the hawkers fell into three broad groups in which there was only a slight variation between the Hong Kong and Hong Kong–Kowloon surveys. First, almost half the hawkers had been resident in Hong Kong since before 1945. An almost equal proportion had spent between 15 and 25 years in Hong Kong, many of whom came as part of the large refugee influx which characterized the 1950s. Finally, a small proportion (5 per cent) had been in the Colony less than 10 years. Different groupings under the heading of 'years in Hong Kong' make comparison with the other surveys difficult, but the survey of resettlement hawkers recorded a much lower percentage of hawkers resident for more than 24 years (22 per cent) and a larger proportion resident since 1945 (70 per cent) which probably reflects the fact that many of these hawkers were squatters who established their homes in the 1940s and 1950s and who had been moved into the resettlement estates during the process of squatter relocation.

Thus, the demographic data on the hawker operatives indicates that they are an old, well-established population, familiar with Hong Kong, with close family and dialect ties. This demographic framework provides the necessary background to a consideration of the hawkers' occupation for, as we shall see, most hawkers have a strong sense of vocation, a sense of the value of their traditional occupation, despite the alternative attractions of employment that have occurred in Hong Kong.

SOCIAL ASPECTS: HAWKING AS A PROFESSION

The prevailing conception of hawkers within the government departments directly concerned with them is that they are, in general, an opportunistic group who take advantage of situations in which authority is too weak to establish itself, and that their business practices cause unfair competition with shops and raise many problems of control.[4] The information of the Hong Kong survey and the resettlement survey is thus of considerable interest in this respect, for these two surveys probed into the number of years of hawking, the reasons for entering the business and the occupational background of the hawkers, and provided a contrasting picture of hawkers in the two areas. It is also important in establishing how employment in petty capitalist activity is related to employment in large-scale capitalist activity.

In an earlier section we indicated that the majority of hawkers were concentrated in either the densely populated tenement areas of Hong Kong and Kowloon or in the resettlement estates. The following data provide a valuable insight into the backgrounds of the hawkers in these two areas.

Table 8.8 shows that there are major differences in the period of time engaged in hawking as shown in the results of the resettlement estates survey, the broader Hong Kong–Kowloon survey and the Hong Kong survey. The majority of Hong Kong hawkers had been in the profession for more than 10 years (72 per cent), a majority of resettlement hawkers for less than 14 years (77 per cent), and of these a sizeable percentage (42 per cent) for less than 6 years. The total Hong Kong–Kowloon hawker population shows a rather more even spread, with some 36 per cent engaged in hawking for 10 years and less, 48 per cent from 11 to 25 years, and only 16 per cent for 26 years or more.

Table 8.8 Hawker operatives by years engaged in hawking (%)

Hong Kong survey		*Resettlement survey*		*Hong Kong–Kowloon survey*	
Years	%	*Years*	%	*Years*	%
< 4 years	13	< 5 years	42	< 5 years	17
5–9 years	13	6–14 years	35	6–10 years	19
10–14 years	26			11–15 years	19
15–19 years	14	15–19 years	8	16–20 years	15
20–24 years	18			21–25 years	14
25–29 years	5	20–23 years	8	26–30 years	8
>30 years	9	>24 years	5	>31 years	8
No information	2	No information	1	No information	1
Total	100	*Total*	99	*Total*	100

There is a clear distinction, then, between the comparatively recently established hawkers of the resettlement estates, and the longer established hawkers of the tenement areas. The reasons for this divergence can partially be explained by examining the occupational backgrounds and reasons for becoming hawkers, and their answers are shown in Table 8.9.

Personal reasons for hawking included such factors as old age, disability and poor health; economic reasons included 'to be my own boss', 'to earn better income', 'can't find another job', 'do not possess skills for any other job' and 'do not need much capital'; social reasons included 'inheritance' and 'friends and relatives in the same business'. The very high proportion of 'no answer given' (recorded as 'no information' in Table 8.9) in the resettlement survey clearly

Table 8.9 Hawker operatives by reasons for becoming hawkers (%)

	Hong Kong survey %	*Resettlement survey* %
Personal	10	14
Economic	40	40
Social	19	—
Others	24	9
No information	7	37
Total	100	100

affects the distribution of responses, but if it is assumed that many of the answers might fit into the social category then rather similar patterns result. The large group of 'others' in the Hong Kong survey includes certain responses which really reflect the hawker's attitude towards business in the particular areas surveyed, including such responses as 'other hawkers were here' which might reflect both economic and social motivation.

Given the rather 'soft' character of the data, it is still interesting to see the limited number of responses that were strictly economic, suggesting that the motivations for hawking were often a mixture of social, personal and economic motivations. A closer analysis of the data from both surveys, by sex and age, suggests that there were some significant variations. Thus, older females above 50 gave personal factors as an important reason for taking up hawking, while economic reasons were more important for the males in the younger age groups. Lack of education was another factor that was closely related to the decision to become a hawker. In fact a high proportion of the hawkers, compared to the total population of Hong Kong in 1971, claimed to have had no schooling at all, and many had had only limited private tutoring which gave them minimal literacy. While this lack of education certainly would prevent their entering 'white collar' jobs, it does not mean that they could not become part of the industrial labour force which has had many employment opportunities.

Another source of information which gives some clues as to the reason for becoming hawkers relates to their previous occupations. Different questions were asked with respect to this question in the Hong Kong and resettlement surveys, making direct comparison difficult. In the Hong Kong survey the hawkers were asked if they were engaged in another job before hawking, and 35 per cent responded that they had been employed before becoming hawkers.

In the resettlement survey the hawkers were asked about their previous employment, but no effort was made to distinguish hawking as one of the previous occupations. This response was grouped together with the 'no answer' category in the tables preventing comparison with the Hong Kong survey on this point. It would be interesting to know if the considerable burst of hawking activity which occurred in the resettlement estates when they were established was simply a transfer of occupation from the squatter areas, or whether it represented a seizure of the retailing sector by squatters who had previously been occupied in other jobs in their squatter locations. The length of period in hawking would seem to suggest that it was the latter, and this is to some extent reinforced by the data on previous occupations.

The previous occupations of the Hong Kong hawkers show a most interesting variation from that of hawkers in the resettlement estates. Thus, of the 224 hawkers previously employed in other occupations 172 (77 per cent) had been engaged in four distinct employment categories. The largest group were former shop salesmen or small shopowners; a second group was made up of former service workers, such as cooks and barbers; third, there were the hawkers who had been factory workers; and finally, a group of unskilled labourers who had not been engaged in factories. The remainder of previous occupations were mixed, but generally came from within the lower-status, unskilled jobs.

The data suggest that the Hong Kong hawkers who have entered the profession from other jobs fall into two categories – those who come as entrepreneurs from other forms of employment, who come from the service sector; and those who have moved from unskilled jobs in factories or elsewhere. Interviews indicate that this latter group were often displaced from these jobs because of their age or disability rather than because of a desire to become hawkers. But most Hong Kong hawkers have been in the profession for a lifetime.

The information on the previous employment of the resettlement hawkers was less specific. The majority (74 per cent) recorded their previous occupation as unskilled labourers or manual workers. Only a small percentage (5 per cent) indicated that they had worked in sales or similar occupations, while 16 per cent had had minor clerical or domestic occupations. This suggests that many of the resettlement hawkers had been engaged in low-paid menial occupations and had taken the opportunity to shift into the resettlement estates in order to establish themselves in hawking. Newspapers of the time provide graphic descriptions of how hawkers had invaded public spaces in the areas surrounding the resettlement blocks even before the people were shifted into their accommodation. Sometimes they even painted out squares on the ground and put in numbers preparatory to taking over the area. This is reminiscent of the squatter invasions of public land in the cities of Peru described by Mangin (1970).

There was also evidence that these invasions were actually organized by groups of people and that prior agreement had been reached on the location of certain pitches. Thus, the evidence on the background of the hawkers would seem to suggest that while the profession has been a lifetime one for most Hong Kong hawkers, there is a sizeable group, both within Hong Kong and in the resettlement estates, who had entered the occupation in an attempt to start their own business. They have behaved, as government officials have commented, in an opportunistic fashion.

Finally, we turn to the question of the relationships among the hawkers. Hawkers have close social relationships spreading outward from the family, to the marketing chains, to dialect communities and also to the hawker associations. Many of these associations appear to combine the qualities of a trade union and a society. First, there are the legal trade unions such as the Hong Kong and Kowloon Licensed Hawkers' Association, whose leaders meet with government officials to try to influence government decisions. Not all licensed hawkers belong to such associations, but, at times of crisis, they can receive wide support from the hawkers.

Second, there are the associations formed by particular types of commodity sellers. For instance, there is the Hong Kong and Kowloon Licensed 'Pak For' Hawkers' Association for hawkers selling clothing and general supplies. The illegal hawkers do not appear to have similar public associations, for obvious reasons, although they often pay small amounts of money to secret societies for 'protection'.

There is evidence that, at the local level, hawkers often rely upon one another for such tasks as minding the stall for a short period. But there is no strong evidence that these hawker agglomerations have formalized associations representing the entire concentration. Rather, hawkers operate in clusters of relatives or friends, often selling the same commodities, and they do not always know hawkers from other parts of the same agglomeration. Only when their existence is threatened by a relocation move do they unite, in much the same way as squatters, in an effort to prevent this happening.

Conclusion: conservation-preservation in the two-circuit system

In the first section of this chapter we argued that understanding the two-circuit structure of cities is helpful in explaining the patterns of economic activity, land use and urban growth. This two-circuit structure is not a static phenomenon. In conditions of rapid economic growth the upper circuit, with its greater productivity, rapidly supersedes the lower circuit, depleting its labour supply and its share of the total distributive and productive activity of the city. But in conditions of comparatively slow economic growth, the lower circuit, with its labour-intensive, low-productive patterns of activity, controls a major part of the city's space and economic activity, although its contribution to the gross product of the city may still be smaller than that of the upper circuit. In the latter condition, particularly if the city is not a closed demographic system, that is, migrants in large numbers are free to move into the city, the lower circuit proves to be highly absorptive with the capacity to create many low-productive occupations.

The study of Hong Kong, however, indicates that even in phases of rapid economic growth, in the 1960s, the growth of the upper circuit at the expense of the lower circuit is by no means evenly spread throughout the various sectors of the city's economic structure. The 1971 Census showed that a much higher proportion (77 per cent) of the labour force in manufacturing is classified as permanent in-workers (that is, workers for wages largely within factory units) than is the case in commerce, where the figure is 59 per cent.

In addition, 29 per cent of commercial labour is listed as self-employed or as various types of family labour while the figure is only 10 per cent for manufacturing.[5] This is a pattern which appears to be borne out by the 1976 industrial survey figures. Although the degree of family employment in both the manufacturing and the commercial sectors is probably underestimated, it indicates that manufacturing is experiencing a much more rapid shift to upper circuit forms of economic organization than is commerce. This pattern of differential rates of growth between the various sectors is, of course, not unique to Hong Kong, and has been noted in the growth patterns of developed western countries (see Bechhoffer and Elliot 1968).

A second theme was the argument that hawkers may be regarded as participants in the lower circuit because of features of their economic organization, social relationships and ecological behaviour. The case study of Hong Kong hawkers reveals that, while a small group is essentially organized like upper circuit firms, the majority of hawkers are, in fact, part of lower circuit activities.

Unfortunately, the survey data were particularly weak with respect to information on the forms of employment in which other members of the family were engaged. While most of the hawker operatives worked full-time, there was conflicting evidence on the proportion of other members of the family employed as hawkers. Thus, 54 per cent of the Hong Kong hawkers reported members of their families working as hawkers, compared to only 19 per cent of the resettlement hawkers. This discrepancy might be partially explained by variations between the two questions asked in the two surveys, which may have been interpreted as meaning either members of the family working in the same business as the operative, or in another hawker business. Also, it may suggest a difference in the pattern of employment among the Hong Kong hawkers, where the operation often involves several members of the family helping out at various times, and the resettlement hawkers, where other members of the family may be employed in factories and help in the hawker unit only occasionally.

If a dual pattern of employment prevails within the hawker sector, with some members of the family working as hawkers and others in different occupations but occasionally helping with the stall, this raises an interesting series of questions on the relationship between the two circuits, not only in Hong Kong but in other Third World cities as well. It suggests that both sectors benefit in terms of labour input which benefits total household income. A situation has been created in Hong Kong whereby the hawkers, as participants in the lower circuit, have shown a remarkable persistence at a time of rapid capitalist expansion.

This phenomenon is explained partly by a structural rigidity in the lower circuit distributive sector which has not collapsed in the face of the growth of the upper circuit, and partly by the fact that the upper circuit units have been slow to perceive opportunities for penetrating the distributive sector of the lower circuit. This structural rigidity is partly associated with the hawkers' spatial inertia: once established in a location, they have proved very resistant to efforts to remove them. Sometimes the hawkers' efforts to establish themselves in a prime location have been thwarted by government. On the other hand, the resettlement estate hawkers have been successful in establishing and persisting in their locations.

Other more specific factors may be cited to explain this persistence. The information presented earlier in the chapter showed that most were over 40 years of age and had entered the occupation before the 1960s, at times when employment opportunities in other sectors were limited. The exceptions to this pattern were those in the resettlement estates, most of whom had taken up hawking during the 1960s.

We have hypothesized that this exception may be explained by three factors: (a) the transfer of hawker occupations from squatter areas to the resettlement areas; (b) the perception of hawking as a rewarding profession in the resettlement milieu, because of inadequate retailing facilities; and (c) the adoption of hawking to replace either illegal occupations, which had been carried out in squatter areas, or occupations (for instance, some types of cottage industry), which could not be transferred to the resettlement estates. Possibly other factors, such as displacement from jobs at the time of the 1966 and 1967 disturbances, have also had an influence, but this does not seem to have been very important in the case of the Hong Kong hawkers.

With the exception of the resettlement estates, the historical pattern has been that the most hawkers have taken up this occupation at times of limited employment opportunities in other sectors. Once established they have found an occupational niche in which they have persisted. Their generally advanced age has meant that, at a time of increasing employment opportunities in the manufacturing sector (which have tended to be directed towards the young and females), they have not been able to pursue this avenue of employment. Similarly, their low level of education has reduced upward mobility into the lower white-collar sector. Thus, they have become occupationally stranded, unable to move out of the hawking profession.

Another factor encouraging the persistence of hawking has been the attitude of the industrial capitalists of Hong Kong towards them as a source of potential labour, at a time of growing labour shortages in the factories in the 1960s. This has been described as a labour-deficit situation, but it was more of a sectoral phenomenon, for the factory demand was for a specialized form of labour, particularly females in the younger age groups. Towards the end of the 1960s this shortage of labour began to extend into other sectors, such as the construction industry. It is interesting to note that in such a situation in other countries, such as Japan, industrialists increasingly invested in a capital-intensive technology, but in Hong Kong there was a reluctance to do so at this point. It may be suggested that this was because of doubts about the political future of Hong Kong which were encouraged by the civil disturbances of 1967.

With this sectoral labour shortage, the hawkers, participants in the traditional bazaar sector, were seen as a potential source of labour for the factories. The thrust of this argument can be traced both in newspapers and in official pronouncements. The issue was first raised in the Urban Council by Mr Hilton Cheong-leen in a June 1964 meeting during which he asked, 'whether in view of the current shortage of manpower in industry, the government is prepared to investigate what concrete steps can be taken to train and redivert unlicensed and licensed hawkers into industry?'[6] The Hong Kong press and debates of the Urban Council at this time are full of complaints that hawkers are unproductive and should be directed into industrial employment, but the hawkers were generally too old and not attracted by wages offered in the factories. Sometimes they had even been dismissed by factories because they had become redundant.

Another aspect of this persistence is found in the small but significant groups who appear to have become hawkers because of the anticipated economic opportunities. These groups usually sell textiles and clothing (often products of Hong Kong's industrial revolution), and seem associated more with entrepreneurial patterns of behaviour rather than those of the majority of hawkers. The case of the Kai Chiu hawkers is a good example of this pattern, and it has also been observed in the night markets. Whether these hawkers are really representatives of firms located in the upper circuit who perceive the market possibilities in the hawker selling outlets cannot be proved, but there is some evidence to point in this direction. If, indeed, this is the case then we have an interesting contradiction, with the upper circuit using lower circuit outlets to expand its operation.

The hawkers' own attitudes towards their occupation may be an additional important factor in its persistence. Interviews with hawkers revealed that while they were inclined to deprecate their profession, reflecting their feelings of marginality in the Hong Kong situation, their reasons for turning to hawking included such explanations as 'independence', 'being my own boss' or 'working with members of the family or friends', which indicate some of the advantages. In addition, they showed considerable pride in their profession, particularly with respect to relationships with customers and knowledge of bargaining procedures, which supports the view that they find the job satisfying.

The general inadequacy of government policy designed to reduce the numbers of hawkers and limit their operations may be cited as an important factor in their persistence. Other reasons for the hawkers' persistence undoubtedly reflect the manner in which they contribute to the city's economy, thus reducing the social overhead capital needed to be invested by both city government and upper circuit employees. They provide goods at cheaper prices than retail stores because they have cheaper overheads. This, it may be argued, is an important factor lowering the general cost of living, and presumably, allowing the upper circuit firms to pay lower wages, thus enabling them to earn higher profits. The hawkers in Hong Kong provide an important outlet for the products of the manufacturing sector, or 'broken cargoes' from shippers.

In many other cities of the Third World hawkers also play a major role in transporting produce from the agricultural sector to the city, breaking the bulk goods down and distributing them throughout the city, but in Hong Kong this task is carried out at different levels by other intermediaries. Hawkers provide a service to the public in that they seek out locations which are convenient to their customers, cutting down the time and cost of travel. This is a major factor in the case of low-order foodstuffs such as vegetables. Hawkers provide an important function in the processing of items in many cities, although, with the exception of cooked foodstuffs, this is not an important factor in Hong Kong.

Hawkers perform an important social function in providing an economic cushion with possibilities of jobs at times of job scarcity. Even at times of employment abundance hawkers contribute to general productivity by utilizing labour which otherwise would not be employed, particularly that of housewives. It may also be argued, perhaps with less evidence, that hawking provides entrepreneurial training which will help economic growth.

Finally, hawkers add an atmosphere to the cities of the Third World which has largely disappeared from the cities of the west. This is particularly true of the bazaars. The *pasar malam* (night market) of Singapore is a tourist attraction, and the floating market is a high-spot for visitors to Bangkok. The tourist potential of certain types of hawker agglomerations appears to be considerable and offers further possibilities of increasing city income.

Despite these obvious contributions that the hawkers make to a city's economy, there is some debate over whether such activities actually promote economic growth. Some writers have argued that the considerable subdivision of activity in the distributive sector creates a barrier to economic growth, since it prevents the accumulation of capital to invest in other activities (see Mintz 1971, 250; Lewis 1958, 400–9; Geertz 1963, 11; Kaplan 1965, 92). While this assertion may have some accuracy (although some of the more successful hawkers in Hong Kong clearly have accumulated capital, if measured only by their stock values), this does not necessarily inhibit economic growth, particularly if the upper circuit in which capital is being accumulated is investing in the cities. It would certainly be difficult to construct an argument that the hawkers in Hong Kong have inhibited the economic growth of the Colony. Indeed, much of the thrust of this chapter is entirely the reverse.

On the other hand, the basis of most urban governments' complaints about hawkers is the argument that they physically inhibit the development of 'modern cities'. The congestion they cause in streets interrupting traffic flow, the illegal sales of raw foodstuffs (posing problems of health control) and their disorderly locations, are all often cited and used as justifications for government actions. In other words, the state is hindered in introducing the elements of collective consumption (roads, etc.) which will make the city more efficient for large-scale capitalism. It is almost impossible to calculate how this aspect of hawkers' operations affects the efficient operation of the upper circuit, thus presumably reducing productivity; but it is hard to imagine that these activities are such a threat to this sector that the economic advantages that the hawkers offer can be discounted. If hawking did prove so attractive an occupation that it prevented entry of its labour force into the upper circuit, then it might be argued that it prevents economic growth, but in the Hong Kong case the persistence of hawkers appears to be explained more by early entrance and occupational stranding than unwillingness to enter the industrial labour force.

Such arguments are, of course, greatly affected by the growth phase of the Third World city. Returning to the two-circuit model of the Third World city elaborated earlier there seem to be very strong arguments against introducing any urban policies designed to eliminate the activities of the lower circuit at the phase of a city's growth when it is dominated by that circuit in population terms.

Small-scale programmes aimed at eliminating problems, such as checking on foodstuff sales to control disease, might be possible; but grand programmes of hawker elimination are almost certainly doomed to failure, for the economic structure of the city is not yet ready to accommodate these changes.

In a sense, this is advocating the approach which J. F. C. Turner (1968) and others have put forward with respect to squatter housing in many of the Third World cities. It is argued that this type of housing will be upgraded by its inhabitants and generally add to the capital stock of the city. Similarly, for the reasons we have elaborated, hawkers will also aid the economic growth of the city by participating in a situation which could lead to a growth in the upper circuit; as for instance, by contributing to lower costs of living in the city. It may be argued that cities such as Calcutta and Jakarta are at this phase of development.

When the upper circuit begins to expand rapidly, the major conflicts between the two circuits begin to emerge, particularly with respect to competition for city space and markets. At this phase, if the Hong Kong experience is valid as an example for other cities, hawkers and other lower circuit activities are perceived by the planners and policy makers of these cities as posing blocks to the growth of the upper circuit. Often this situation is presented in the language of the traditional-modern dichotomy: traditional persistences are blocking the growth of the modern sector. Hawkers get in the way of motor cars or public transport. The land on which squatters are located is needed for industrial expansion or public works development. The inadequate waste-disposal systems of cottage industry are said to be polluting the city.

All this is, in a sense, quite true, but the Hong Kong experience suggests that these blocks do not necessarily prevent economic growth, or for that matter slow down the pace of modernization of the city. Indeed, the links between the upper and lower circuits and the manner in which they complement each other may help the process of overall growth by preventing structural unemployment in the labour force from emerging and by allowing a generally higher level of total labour input to be continued. However, at this phase, the lower circuit activities often impel city administrators to move much more rigorously against them, with schemes of relocation and resettlement which severely limit the activity of the lower circuit. Clearly, such programmes have to be carefully devised in relation to the pace at which the upper circuit is developing.

Finally, it may be argued that the Third World city will pass into a phase when the upper circuit dominates and the lower circuit activities wither away, remaining only as traditional remnants that persist in the interstices of the economy. It can be argued with some force that by 1980 Hong Kong and Singapore were moving very rapidly towards this stage and that the elements that are part of this process are creating conditions which will see a rapid phasing out of the hawker's role. The ecological conditions in which the hawker flourishes, while persisting in the resettlement estates, are showing signs of diminishing. The high density tenement core of the inner parts of the Hong Kong–Kowloon metropolitan area has decentralized with the concomitant growth of the outer parts of the city.

There is some evidence that the upper circuit firms have begun to try to control the distributive sectors previously controlled by the hawkers. This process is aided by the growth in household income in Hong Kong, which has led to an increase in the level of consumer ownership of items such as refrigerators. Attitudes toward food purchase may also be changing slightly, with an increase in the consumption of instant food which is more western in character. The movement of supermarket complexes into the housing authority estates has already begun and is increasing rapidly.

The growth of a more educated labour force and the extension of possibilities of employment in other sectors, especially for the hawkers' children, may mean that they will not themselves become hawkers. Of course, this point is difficult to measure, for it rests upon developments in the labour force as a whole which are not easy to predict. A reversal in economic conditions, particularly in the manufacturing and investment sector, might create a situation in which the population once again turns to hawking. This appears to have occurred in Hong Kong between 1978 and 1982.

Finally, government policy towards hawkers now seems more designed to attack the structural aspects of their operation, including preventing inheritance and increasing licence payments. Of course, the success of this policy rests upon control and enforcement, but the macro-conditions in the urban areas would seem to favour its success. Thus as Hong Kong moves towards the 1990s, the hawkers will gradually begin to lose control of the distributive sector, and their picturesque occupation which has contributed so much to the Colony's economy will eventually fade away. They will have served their purpose for the capitalist mode of production and can be largely discarded.

9

Women workers or working women? A case study of female workers in Malaysia

> The usefulness and values of most things depend, not so much on their own nature as upon the number of people who can be persuaded to use them.
>
> (*An Advertiser's Guide to Publicity*, 1887)

Introduction

For the last few years the weekly journal *Asiaweek* has sponsored a short story competition for Asian writers which has produced some excellent entries. The runner-up in 1982 was a short story entitled 'Mr Tang's Girls', by Shirley Geok-lin Lim, a Malaysian.[1] The plot of the story revolves around Mr Tang's second family (wife, four daughters and *amah*) whom he maintains in Malaka, visiting them every weekend. His first family lives in Kuala Lumpur where Mr Tang is a successful (or perhaps moderately successful) businessman engaged in tin mining. This appears to be a comfortable arrangement for all.

The dramatic intensity of the story develops around the efforts of Mr Tang to arrange a marriage for his eldest daughter in the Malaka *ménage* with one of his employees. While seemingly accepting this on the surface, the decision eventually drives the daughter mad and she kills her father. Starkly presented, the story seems to involve little that is new. The now rather hackneyed theme of the clash between tradition and modernity that is epitomized by arranged marriages is common in Asian literature. But it is the underlying theme of the story that is of considerable interest. Mr Tang's eldest daughters live in a world that is full of dreams of fashion and beauty. Let us illustrate this world.

> Kim Mee had painted her toenails a new dark-red colour; she was going to a picnic in Tangjong Bedarah on Sunday and wanted to see the effect of the fresh colour on her feet bare on sand. . . .
>
> But there was Saturday night and the evening meal late at nine and the girls would sit in the living room with long washed hair reading *Her World* and *Seventeen*, selecting patterns for their new frocks.

The eldest daughter's growing insanity brought on by the trauma of the arranged marriage is strikingly shown by the use of cosmetics.

> When he came back next Friday, Kim Li had gone through a total change. 'I'm a woman now', she had said to her sisters and began using Kim Mee's make-up every day. She pencilled her eyebrows crudely, rubbed two large red patches on her cheeks and drew in wide lips with the brightest crimson lipstick in Kim Mee's collection. After every meal she went to her room and added more colour. Blue shadow circled her eyes, and her clumsy application of the mascara stick left blotches below her lids like black tear stains. She teased her short hair into a bush of knots and sprayed cologne until it dropped down her neck.

This lengthy quotation is cited not for its dramatic effect, but to show how in the increasingly insane mind of Kim Li the cosmetics (beauty) bestow on her the right to womanhood, and also illustrate how the need for cosmetics has become part of her dreams and fantasies.

Another theme less developed in this short story is the idea of romantic love. Kim Li's initial reaction to the announcement of the arranged marriage is to accept it, but then she asks her father to arrange a date so that she may meet her proposed husband. Her father responds that this is not permitted: 'You mustn't spoil the match by acting in a western manner.' But this theme is not developed any further in the story. Other writers have shown how values such as romantic love are transmitted through movies, television, popular music and journals (see K. Robinson 1985).

It would seem that in the realm of ideas, fantasy and dreams, there are processes occuring that are shaping young women's attitudes in Asia today. These are not simply summed up by conventional theoretical frameworks such as traditional versus modern or westernization. They are much more complex than these models would allow, and need careful evaluation.

This chapter discusses the process of proletarianization (narrowly defined as entry into factory wage labour) arguing that this process is much more complicated than many existing theoretical models suggest, and in particular that the creation of consumption needs plays an important role in inducing non-proletarians into the labour force.[2]

Proletarianization, export zones and women workers

The last few years have seen the emergence of a considerable number of studies that describe the development of export processing zones in Asia.[3] These studies fall into three broad categories. First, the macro-studies, which analyse the emergence of the Free Export Zones at the level of the international economy, variously described as the 'international division of labour' or 'multinational diversification of off-shore investment into cheap labour regions' (see Fröbel, Heinrichs and Kreye 1980; Keesing 1979; Nayyar 1978). Since this process has been documented in Chapter 5 the findings are not repeated here. A second thrust of these writings has focused upon the conditions and features of the work force of these Free Export Zones, documenting the fact that the labour force is often female, from rural areas, and concentrated in the age groups between 16 and 30.[4] Finally, a smaller number of studies have begun to focus upon the lives of workers in these zones at a micro level, utilizing detailed ethnographic studies (for example, Salaff 1981).

While figures vary it is generally accepted that women make up to 70 per cent of the employment in many of these industries, particularly when they are located in the Free Export Zones which have been established to facilitate this type of industry. The predominance of women in these labour forces has often been the subject of surprised comment, particularly in the societies of Asia where women's family roles have been traditionally emphasized. It is, however, not a new phenomenon for, from the earliest stage of the Industrial Revolution in England and France labour-intensive industries such as garment production, textiles, food processing and, more recently, electronics, have traditionally used female labour.

The explanations for this dominance of women in the workforce vary, and often utilize sexual stereotypes which emphasize the docility, patience, manual dexterity and visual acuity of female labour. Of greater significance, probably, is the fact that female labour is often paid much less than male labour for the same job, and since cheap labour is essential for these labour-intensive industries this is an important point for employers.

Historically, the process of proletarianization has been the consequence of the expansion of capital, and an important mechanism for the dissolution of non-capitalist relations. Safa[5] has shown how this process of labour recruitment operated in the history of the garment industry in the USA, in which she demarcates three stages:

(a) Use of native labour force, including recruitment from the rural area.
(b) Immigrant labour.
(c) The 'runaway shop'.

In the earliest phases of the garment industry in New England the workers were generally daughters of farm families; young, single women who could easily be withdrawn from agricultural labour and usually stopped working once they were married. As Safa comments,

> Women's lives were thus neatly divided into a paid productive phase and an unpaid reproductive phase in which their energies were largely directed toward the maintenance and reproduction of the labour force, which should have helped to assure the maintenance of a steady supply of native labour.
> (Safa 1980)

By the middle of the nineteenth century this source of labour began to contract as alternative sources of female employment became available, and at this point immigrant labour in urban areas replaced the female labour force. For a number of reasons, partly, it has been argued, because of the immigrant ethos of hard work and greater poverty among the new groups, this resulted in even more exploitative conditions, including an increase in 'piecework' production. The final phase is that of the 'runaway shop' which represents an ongoing strategy on the part of transnationals to utilize several trends in the international economy (see Chapter 5) which favour this new international division of labour.

This relocation of labour-intensive industries raises important questions as to what impact it is having on Third World women in the areas in which new industry has moved. What are the forces that are leading to a dissolution of pre-capitalist labour relations in the countryside? Are the new sources of employment provided for these women improving their socio-economic position in the society? How are the household relations being restructured as a result of these changes? Is wage labour increasing the class consciousness of these working women by removing them from the home and exposing them to the wage labour markets? Most significantly is there any tendency to follow the patterns that have emerged in western countries?

Writing in 1982, Salaff and Wong suggest that there are sharp differences in the participation of women in the labour force in the first Industrial Revolution of the nineteenth century, where women's employment in manufacturing and service occupation such as domestic service was important, and the second 'industrial revolution' of the twentieth century where better educated women were drawn into jobs that demanded literary and numerary and organizational talent (Salaff and Wong 1982, 1). In effect the diverse patterns of the expansion of international capital in collaboration with the states of Asian countries is producing situations in which both these industrial revolutions, previously separated in time, are occurring at the same time. Thus Singapore has already entered the second industrial revolution and therefore the skill demands for women in the labour force are very different from 15 years ago. The remaining labour-intensive industries are now heavily reliant upon contract labour (largely Malay) which is an ironic reversal of the history of the nineteenth century. In Malaysia, on the other hand, industrialization is still demanding cheap labour which is being largely recruited from rural areas. It is this latter situation which is analysed in this chapter.[6]

Female proletarianization and urbanization in Malaysia

Malaysia has experienced three waves of urbanization. The first, from 1785 to 1930, was characterized by the creation and consolidation of the port towns of Penang, Singapore and Malacca, and the growth of mining towns such as Kuala Lumpur and Ipoh. These towns were primarily populated by non-indigenous migrants, while the indigenous Malays were largely rural (see Puthucheary 1960). The second period of urbanization, from 1947 to 1970, was characterized by political turmoil, the resettlement of a large number of Chinese in 'New Villages', the creation of independence and a continued growth of towns in the west coast. The number of Malays in the cities increased but they still remained largely in rural areas. But with the introduction of the New Economic Policy in 1970, aimed at reducing the barriers of a 'plural society' and eliminating 'poverty' among all races, the policy conditions were created for a very radical transformation in Malaysian society.

This is certainly revealed by the dramatic changes that have occurred in urbanization since 1970. There was an increase in the urbanization level from 27 per cent in 1970 to in excess of 40 per cent by 1980 (see Chan and Kok 1984). Of even more importance is the growth of Malays in the urban areas which, as Table 9.1 shows, reveals the fastest rate of urbanization of any ethnic group, at 6.7 per cent per annum compared to 3.7 per cent for the Chinese. In addition the Malays have increased their proportion in the urban areas from 27 to 33 per cent. It is against this backdrop of rapid Malay urbanization that the growth of Malay female employment in factories must be assessed, for this increasing urbanization has been fuelled by a significant industrial revolution.

Since 1960 Malaysia has recorded a significant growth in manufacturing, with the percentage contribution to the Gross Domestic Product increasing from 9 per cent in 1960 to 24 per cent in 1980. Initially, much of this industry was concentrated on 'import substitution' of major products such as automobiles, household

Table 9.1 Peninsular Malaysia: population in urban areas by ethnic groups, 1970–80

Ethnic groups	*1970*				*1980*			
	Urban 000s	*%*	*Total 000s*	*%*	*Urban 000s*	*%*	*Total 000s*	*%*
Malay	713	27	4,822	53	1,359	33	6,384	54
Chinese	1,557	59	3,274	36	2,234	54	4,136	35
Indian	338	13	978	10	508	12	1,239	10
Others	30	1	73	1	47	1	90	1
Total	2,638	100	9,147	100	4,148	100	11,849	100

Source: Salih (1983a)

appliances and some food processing industries. The recruitment of labour tended to concentrate upon surplus pools of labour available in urban areas adjacent to these new industrial areas, which meant that a majority of this labour was Chinese (reflecting the numerical dominance of the Chinese in urban areas) and predominantly male (80 per cent of total employment) (see Wheelwright 1965).

But, after 1970 a number of multinational corporations, seeking cheap female labour, began to establish factories in Malaysia. Among these industries the electronics industry tops the list in terms of rates of growth of factories and number of female employees. In 1970 there were 41 electronics firms employing 3200 workers (99 per cent of whom were women) and by 1976 the number had increased to 138 firms with a total of 47,000 workers. The garment and textile industries show similar patterns.

This impressive growth in the number of transnational-owned establishments is partly the result of the favoured treatment accorded by the Malaysian government. Since 1970 it has embarked upon an ambitious programme of inviting foreign investors to establish in Malaysia. Besides general incentives (export processing zones, pioneer industry, tax breaks) the Malaysian government has also ensured an industrial environment conducive to foreign investors. For instance, legislative regulations prohibiting night-shift work for women workers have been relaxed and efforts have been made to curb the rights of the trade unions. Since 1970 the Malaysian government has sent several trade missions abroad which publicized the 'fast-fingered' Malaysian female workers, and within the country manufacturing firms are allowed to send recruiting agents to the countryside to encourage rural girls to migrate to the factories.

This change of industrial policy after 1970 was an important part of the New Economic Policy (Dasar Economi Baru) introduced after disastrous communal riots in 1969. The New Economic Policy was designed to reduce poverty and structural inequality between ethnic groups in the society. Among other things, it set forth the requirement that the intake of workers should reflect the racial composition of the society, and as such not less than 30 per cent of the workforce in every industry must be Malays.

Since its implementation, reports on the composition of manufacturing firms show that more than 50 per cent of their unskilled workers are Malays and, in the case of firms employing a majority of female workers, most of them are Malay girls from rural areas. The New England experience appears to be being repeated.

Much of this new industry is located in urbanized areas, with over 78 per cent setting up in the more developed urbanized west coast states of Selangor, Perak, Johore and Penang. Since the majority of Malays (85 per cent) still resided in rural areas in 1970, the growth of these industries has involved a steady movement of Malays from rural to urban areas.

Overall, the effects of this new industrialization have been dramatic. Early research on the labour force participation in Malaysia has been carried out by Gavin Jones (1965) which seemed to support the idea of a declining female labour force participation until 1957, except for a slight upturn among younger females (18–24 years) in urban areas. A later paper by Hirschman (1976) comparing 1957 and 1970 census data, indicates growing female participation which is supported by the work of Ariffin (1980, 1981). Table 9.2 shows a remarkable increase (320 per cent) in manufacturing employment between 1957 and 1976. But even more significant has been the increase in female labour (912 per cent), which has seen the proportion of women in the manufacturing labour force grow from 17 per cent in 1957 to 41 per cent in 1976.

In the context of the Malaysian economy, this growth of proletarianization is particularly significant with respect to the growth of employment by race. Table 9.3 shows that the most significant change in the racial composition of the manufacturing labour force has been among the Malays. Even more important is the growth of the Malay female labour force from 8000 to 110,000. This contributes a major part of the shift to a general increase in female labour in the

Table 9.2 Employed males, females in the manufacturing sector of Peninsular Malaysia, 1957–76

	Total	*Male*	*Female*
1957	135,382	112,837 (83%)	22,545 (17%)
1970	251,939	178,881 (71%)	73,058 (29%)
1976	584,341	342,828 (59%)	241,513 (41%)

Source: Ariffin (1981).

Table 9.3 Employees in the manufacturing sector of Peninsular Malaysia by race and sex, 1957–76

	Total	*Malay*			*Chinese*			*Indian*		
		Total	*Male*	*Female*	*Total*	*Male*	*Female*	*Total*	*Male*	*Female*
1957	135,382	26,588	17,699	8,889 (39.4%)	97,502	84,054	13,448 (59.6%)	10,086	9,937	149 (0.7%)
1970	251,939	73,051	45,012	27,949 (38.3%)	164,497	120,758	43,739 (59.9%)	13,254	12,061	1,193 (1.6%)
1976	584,341	210,868	101,097	109,771 (45.5%)	328,985	209,453	119,532 (49.6%)	41,005	29,415	11,590 (4.8%)

Note: 'Others' have been excluded from this table so numbers do not add up exactly.

Source: Ariffin (1981).

manufacturing sector, which has grown from 17 per cent to 41 per cent. In the prevailing racial situation of Malaysia few commentators would have predicted this remarkable increase in Malay female employment.

The factory system

In this section we will discuss the four major features of the dramatic increase in the number of Malay female workers under four headings: recruitment, work and payment system, occupational mobility and the factory socio-cultural system.

RECRUITMENT

The majority of female Malay workers in the factories are young, single women in the age group between 18 and 24 who live with other factory girls in the cities. This represents a new trend in Malay migration because before 1970 the majority of single migration was male, and Malay society's perception of single female migration was negative. If Malay single female migration did occur it was mostly to be employed as domestic servants living with other families.

Most of the workers are drawn from rural areas and many firms have launched recruiting campaigns which emphasize good pay and working environment, and the benefits associated with factory life such as free work uniforms, good medical facilities and beauty-grooming classes. Research conducted by Ariffin (1980) found that the major reason for migrating among these workers from rural areas was the desire to improve their economic positions (74.2 per cent). The analysis of the data relating to their background reveals the rationale behind this motive.

The majority come from poor families whose household heads are small-scale farmers or low-level government employees. Seventy per cent of the households from which these girls come have incomes of below $M250 compared to the national per capita figure of $M2206. The survey also reveals that many of the girls have only middle-level educational achievements, insufficient for entry into anything but unskilled factory labour.

The second most quoted reason for migration is the desire to attain more individual freedom. Nineteen per cent of the sample put this as their major reason. This reflects the encroachment of modern norms into Malay rural villages through mass media, modern-type education, government development and the return of migrants with information on the urban situation, which have undermined the traditional values which kept Malay girls of marriageable age at home.

Other surveys have produced rather similar results, although the HAWA survey of 1066 female workers in Kuala Lumpur and Johore lays much less stress on this second motivation, emphasizing economic need more.[7]

WORK, PAY AND INCENTIVES

Contrary to the common experience of many migrants, most of the female workers found jobs rapidly, the majority (60 per cent) in electronics, 13 per cent in textiles and 13 per cent in garments. In most of the factories the girls are taken on initially as probationary workers for a 3-month period, and after that their average monthly income is between $M100 and $M200 a month. Generally this represents two-thirds of the income earned by male workers.

Work hours vary according to the industry, but in electronics there are three shifts for the full days. Most workers are employed for six shifts a week. Permanent workers are entitled to benefits and participate in off-factory entertainment.

This factory environment creates new situations for these girls. First, the opportunity to mix with men; second, earning money and learning to use it independently; third, some leisure time, and, finally, participating in public functions outside the family. Despite concerns about their work situation, particularly insecurity of jobs, most of these workers are determined to stay in urban areas. Their concern with security is justified, for some companies retrench workers in order to avoid paying bonuses and some are dismissed after just 3 months of employment.

OCCUPATIONAL MOBILITY

The possibilities for women advancing in these factory jobs are limited, for the majority of skilled workers are males. This situation leads to a great deal of movement between factories, as workers try to seek more beneficial conditions or wage rates.

THE FACTORY SOCIO-CULTURAL SETTING

Within Malaysia there are three main types of factories: western, Japanese and Malaysian. In the western factories management encourages their female workers to emulate stereotyped versions of affluent and modern western women, encouraging sporting activities, dating, beauty contests and dances. Japanese companies encourage their workers to be gentle and passive, with the manager portrayed as an authoritarian father-figure. The Malaysian companies, who generally cannot compete with the large transnationals, attempt to emphasize Malay norms with picnics and other activities (see Smith 1982).

The urban setting

The urban factory introduces rural girls to many new norms and lifestyles. But, in addition, they have to adapt to an urban environment with which they are unfamiliar. The majority of companies do not provide housing, and thus the girls stay in cheap housing arrangements in poorer areas. This follows two basic patterns: one, a group of girls rent a house and share expenses; two, they stay with landladies and pay for board and lodging. These housing conditions are scarcely favourable, with a high degree of overcrowding. For example, in a small house consisting of three rooms, it is very common to find fifteen or more occupants. Finally, the amenities are often inadequate, with limited cooking and other arrangements.

Urban living also involves other adjustments for the female workers with increasing freedom and leisure time. In Malaysia the press has focused on such activities as prostitution, drug-taking and illegitimate childbirths, suggesting that this growth of female labour is not without its social costs. There is also increasing evidence of industrial health problems, particularly affecting eyesight.

It is hardly surprising that given the rapidity of the growth of the female labour force the state has been slow in reacting to needs of the urban female workers, particularly those in the export zones. Thus legislation, education, infrastructural changes, and guidance and counselling are generally inadequate to cope with the problems of these workers, and it has been left to the activities of community and non-governmental organizations to try to develop institutional responses for these problems. These inadequacies are most serious in the area of legislation where the government, in its efforts to attract foreign investment, has relaxed laws in these export zones as, for instance, with respect to shift work, labour unions and compensation on dismissal (see Lim Lin Lean 1982).

A Penang case study

This process of proletarianization is particularly well illustrated by the case of Penang State in Malaysia (see Salih 1983b). Prior to 1969 Penang's economic survival had rested largely on its free port status and a variety of small processing activities relating to rubber and tin. With the loss of free port status in 1969 the State Government developed a bold policy of encouraging industrial activity in the State, which led to the establishment of serviced industrial estates offering various kinds of tax incentives and tariff protection for both import-substitution type industry and exported manufactured goods. This policy has been remarkably successful.[8] The manufacturing sector's share of gross domestic product has increased from 21 per cent in 1971 to 37.2 per cent in 1980. This industrial growth fuelled a considerable increase in per capita gross domestic product which in 1971 was $M1035 but by 1980 had grown to $M2357. In addition, employment in the State's manufacturing sector grew from 16 per cent in 1970 to 29 per cent in 1980. Much of the organizational basis for this growth was provided by the Penang Development Corporation (established in July 1970) as the principal development agency of the State Government.

This growth of manufacturing was largely located in two types of industrial estates (see Figure 9.1). In 1982 the first type of industrial estates, which were not Free Trade Zones, employed 16,346 (30 per cent), of which only 6982 workers (42 per cent) were females. The major types of industry in these areas are food processing, chemical fertilizers, metal products and rubber-based industries. Malays make up some 41 per cent of the labour force in these industries.

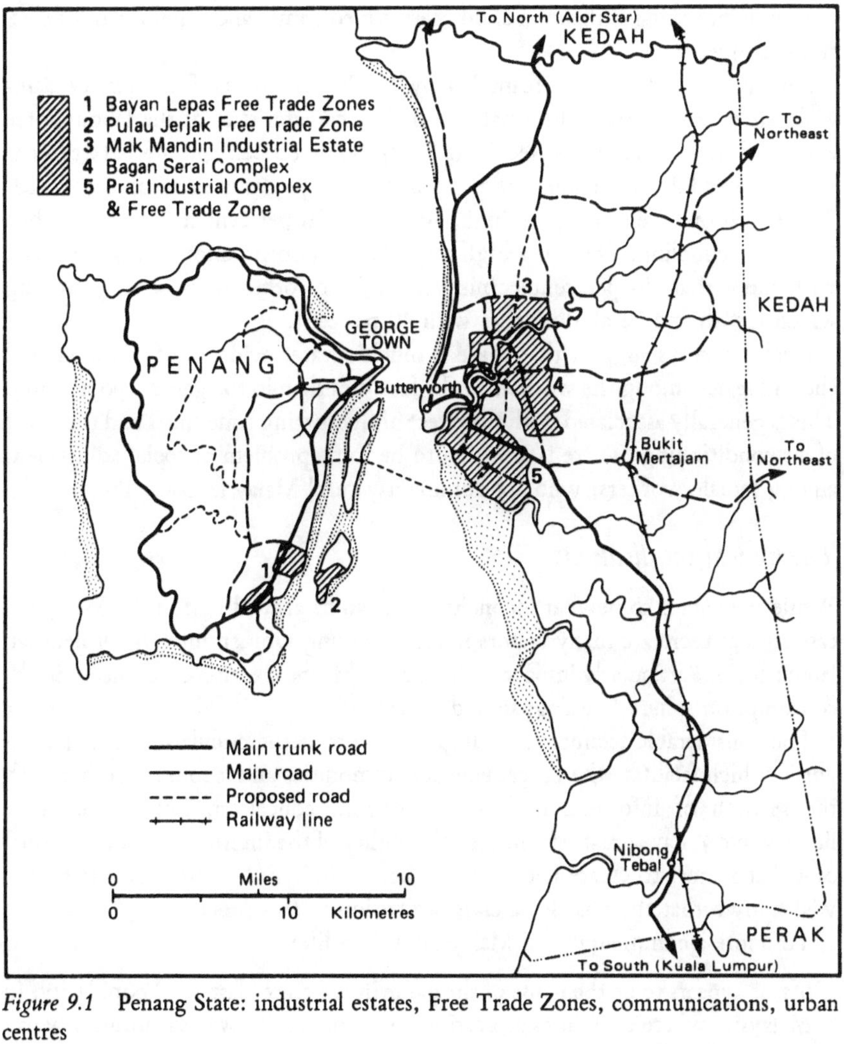

Figure 9.1 Penang State: industrial estates, Free Trade Zones, communications, urban centres

The second type of industrial estates are the Free Trade Zones, of which there are four in Penang State located at Prai, Prai Wharf, Bayan Lepas and Pulau Jerjak. In 1982 Bayan Lepas was by far the largest of these zones with 38 of the 50 factories employing 30,215 of the 38,434 workers. A much higher proportion of female workers (70 per cent) is employed in these FTZs and also a higher percentage of Malays (46 per cent). The two major types of goods produced in these zones are electronics and textiles.

A very high proportion of the factories in these zones are foreign owned. They are branches of international companies producing electronic circuitry (silicon circuits) in Japan and the USA, which are then sent to Penang for assembly and testing before being utilized in computers, video games and other forms of electronic gadgetry.

Surveys conducted by a team led by Dr Kamal Salih of University Sains Malaysia give important information on the characteristics of the labour force employed in these factories (see Salih 1982). Some 83 per cent are between the ages of 16 and 27, with about 38 per cent in the age group between 20 and 23. The education levels are quite high, with over 70 per cent having more than 'Form 1' schooling. There is a high migrant component among these workers, with more than 46 per cent coming from States other than Penang, among which Kedah and Perak dominate with 40 per cent.

Health surveys suggest that there is much greater evidence of intestinal and chest illnesses among the workers in the factories than in the general population. This is generally associated with the effect of fluctuating time shifts and the effect of air conditioning. There is also said to be some problem of social adjustment among female workers, who are dominantly rural Malay females.

Theoretical implications

While the data is somewhat inconclusive it can be argued that in the Malaysian case studies there are many factors inducing young women into the proletarian labour force. Particularly important in our view is the creation of felt needs in the consumption sphere (see Lagbao and Pa 1981).

The considerable technological improvements at the circulation level (transport), which facilitate the movement of commodities and labour, together with the growth of information flows (television, education, improvements in literacy, etc.), have greatly facilitated the ability of the international and national capitalist sectors to create felt needs for people in Third World countries. It is well known that there is also a class dimension to this process.

To quote a commentary on Malaysian urban life:

> Status appears to be the name of the game in the rise of fast-food popularity. In Malaysia, where a car sticker bearing the name of an overseas university can

> open doors, and where office workers plunk down a month's wages to buy a belt with a designer buckle, chomping American burgers and guzzling root beer helps promote the image. (Robinson 1982, 7)

Historically, four crucial components have been necessary to bring about the increase in demand which allows the creation of these mass markets. Hamish Fraser (1981) has shown how these components operated in Britain between 1850 and 1914. First, during this period there was a growing market in terms of numbers of customers. Second, there was a growing amount of disposable income. Third, there was a growing rate of literacy and education on the part of customers. This, together with organizational changes in the structure of retailing, credit, distribution and production facilitated the mass production of commodities for this growing population. Finally, the growing differences between the bourgeoisie and the working classes created an ideological environment in which advertising would utilize status as an important promotion style in creating demand for the mass products.

This last component of the creation of the mass market (mass advertising) was to become much more important in the twentieth century, and was most fully developed in the USA after 1910. The process now is reaching its peak in the world advertising campaigns of the transnationals, and is brilliantly described in Stuart Ewen's book *Captains of Consciousness* (1976).

Underlying this growth of the mass market were two major processive changes that not only facilitated it but were crucial to its emergence. The first was the large increase in numbers of wage workers concentrated in urban centres. The second change was the creation of a new family form, separated from the production process. The emergence of industrial capitalism led to a restructuring of the family as a consumption unit within which new needs began to take shape. The reconstitution of the family did not take place quickly or evenly, but by the twentieth century the family in countries such as the USA and Britain became the major market for mass-produced goods. As Zaretsky points out:

> Mass production forced the capitalist class to cultivate and extend that market, just as it forced it to look abroad for other new markets. . . . Working people now see consumption as an end in itself, rather than as an adjunct to production, and as a primary source of both personal and social (i.e. 'status') identity. (Zaretsky 1976, 68)

While we would not wish to argue that these processes in Asia are identical to those in Britain and the USA, it is surely no coincidence that the countries which have been the major locations for the emergence of the export processing zones

(Republic of Korea, Taiwan, Hong Kong, Singapore, Malaysia and the Phillipines) have also been the countries in which the processes and conditions outlined in the preceding paragraph have been most developed. What we are suggesting here is that the process of proletarianization involves more than just a change in the production relationship – it involves equally the creation of consumption needs and desires. These are obviously not simply manifested at the level of a desire for a particular commodity, but can be part of the money-earning process aimed at increasing access to education in order to enable upward occupational mobility, status and possession of commodities.[9]

Historically, using European examples from the eighteenth century, Scott and Tilly have argued that 'traditional families employed a variety of strategies to promote the well-being of the family unit' (Scott and Tilly 1981, 139), in which off-farm deployment of labour was characteristic. In the nineteenth century therefore, in the face of major structural changes, the deployment of women and children in factory labour, they argue, involved no great change in family strategies of off-farm employment. The model they use 'posits a continuity of traditional values and behaviour in changing circumstances. Old values coexist with and are used by the people to adapt to extensive structural changes' (Scott and Tilly 1981, 131). These arguments seem to be very helpful in explaining why parents allow young single Malay women to take up factory employment.

They are only partially helpful, however, if it can be shown that these single women workers in factories of export zones are returning a large proportion of their incomes to their families. If this is not the case, and income is being used by the factory workers to spend on personal consumption or saving for future contingencies, such as marriage, then the theory of 'value lag' is inadequate. We would argue that the milieu in which young women enter the labour force is more complicated. While 'value lag' may explain some family strategies we also have to accept 'value change' as an important criterion. The creation of consumption needs is certainly an important factor affecting the entry of these young women into the labour force. This is a process accelerated by the technological means to create these consumption needs (television, etc.) which means that the dreams of young women workers are no less important than family strategies.

Of course, changes at the production level in rural areas (increased mechanization, etc.) are also creating a situation in which female labour is required less, and thus there is an expulsion process occurring at the production level. Equally, the process could not be occurring if there were not jobs being created in the export zones as a consequence of the 'new international division of labour'.

Of course, these questions are only speculative and need more research, but certainly an understanding of the complexity of the whole process of labour recruitment is crucial to the investigation of the formation of a stabilized proletariat located in urban areas.[10] If these young women factory workers are responding to a family strategy of increasing incomes, then their tenure in the workforce will be greatly affected by the point at which their family desires them to leave the workforce. If, on the other hand, they are motivated by a desire to increase individual consumption and saving, the decision will rest on their own shoulders. Of course it may be that the specific requirements of the factory labour force (age, manual dexterity, etc.) will lead to their expulsion, but this is a factor over which they have little control. There is much debate over the question of the stability of the labour force in these export zones. Some workers comment on the 'footloose' character of these industries, suggesting that they will relocate in the face of increasing wage levels,[11] but in the case of Malaysia (particularly the electronics industry) there appears to be considerable stability, and indeed many firms are expanding plants.

Conclusion

Of course it may well be that we are forcing the role of the consumption sphere as too dominant a reason for the dislocation of females from non-capitalist systems of production; certainly research into these questions will also need to explore the production and circulation spheres. But, ultimately, the understanding of the consumption needs of 'Mr Tang's girls' may well be crucial to the broader understanding of the growth in the numbers of women workers in Asia and the formation of a female proletariat.

10 Conclusion

One issue is basic to the thesis we have been developing in this study: it is that of the relationship between the pace and scope of international capitalist expansion and the central role that city systems play in this process.

Writing seventeen years ago, we had not foreseen just how dynamic the internationalization of capital would be, and, indeed, argued that the gradual expansion of capitalism in most economies of the Third World, together with the time bomb of population increase, was bringing about a condition of rural and urban involution in many of these societies. From our perspective then, the important research questions centred on the ability of Third World countries to 'handle' these involutionary situations over the medium term. It seemed to us that urban involution would produce the most logical temporary resolution for these contradictions, but in the long run such contradictions would need to be resolved in other ways: through a widening of capitalist relations of production or by social revolution.

The principal factor that has operated to speed the first of these options is the introduction of production capital on a large scale into the Third World by transnational corporations and the facilitating of this movement by the promotional strategies adopted by many Third World states. Governments have encouraged liaisons between international capital and their own local capitalists in import-substitution manufacturing activities; but, especially since the 1970s, they have encouraged direct investment by transnationals acting alone, in production for export markets.

Of course these processes could not have occurred as they did without changes in the social structures of Third World countries, since they largely rested, at least initially, upon the production of consumer goods for the upper- and middle-income classes which were, and are, mainly located in the urban centres. At the same time, the penetration of capitalism into the countryside through the expansion of cash cropping, new agricultural production technologies, and the growing proletarianization of the peasantry dislocated rural populations who migrated to the cities or became full- or part-time wage-labour for the export-oriented production of agribusiness and local producers.

Our respective research experiences over the years in providing first-hand knowledge of these processes, have made us realize that they must be understood within the context of a study of social class and capital. Our work here has therefore analysed these processes and relationships at various levels: first, at the level of the international capitalist system and the different theoretical perspectives put forward to explain its operation; second, at the level of the national society in which the interaction of the state, national and transnational capital is central; and, finally, at the level of the region and the workplace, where the immediate human impact of these processes can best be comprehended.

Cutting across these spatial, sectoral and social categories are the urban systems which stretch from the great global metropolises of New York, Tokyo and London, through the continental centres such as Singapore and São Paulo, to the national urban hierarchies, from primate cities and central sub-systems to the smallest market towns. One point in particular has been stressed: that such cities are the principal theatres of capital accumulation and class interaction in the societies of Asia and Latin America.

The upper levels of these urban networks are the *loci operandi* for transnational corporations and the local ruling classes, for, as Cardoso (1975) has pointed out, it is in the large urban marketplace that capital can most profitably function. It is thus within the context of the city system that most capital accumulation occurs, and where the system can prepare for each new phase of expansion. Needless to say, not all the surplus appropriated from society's production is generated in the cities. For that reason we have had to look beyond the urban areas to see how the system of centralized accumulation works. So we have examined the ways in which the surplus produced by peasants, craftworkers and agricultural labourers in the rural areas and small towns is channelled away into the large cities to add to the surplus extracted from those in the modern capitalist urban sector, as well as, for instance, from hawkers in the lower circuit.

While the study of the technical mechanisms and spatial consequences of the tendency towards concentration in the upper level of the urban system is important, we have focused particularly upon the patterns of social change which arise from these processes. This has been a study of Third World urban social structures and class relationships, and of the interaction between such internal structures and external forces. The Third World metropolis has become the central place for social co-operation – and confrontation – and for the central activity of capitalist development.

Our objective has been to look at the question of development/underdevelopment through the prism of capital accumulation. This has provided a perspective which allows us to reinterpret certain concepts which have appeared in development literature over the past three decades. For example, 'capital formation' in reality involves the extraction and appropriation of the independent small producer's surplus; 'modernization' becomes a development strategy geared to promote corporate (both state and private) accumulation and production, and a way of opening consumer markets to the products of transnationals and national oligopolies.

Furthermore, the very mechanisms which introduce new technology and larger-scale organization into industry and agriculture deprive the producers of their land, their craft production and their employment, transforming them fully or partially into wage labourers for capitalist production. Similarly, the loans and other financial transfers required for such technology in the modernized sectors create an indebtedness which locks Third World countries further into international system dependency. This is facilitated by rapidly accelerating communication and information flows which are dominated by the industrialized countries. State involvement in support of this strategy more often than not uses public funds to subsidize and facilitate the operations of corporate capital, even in apparently socially oriented projects such as urban shelter for slum dwellers (Payer 1982, Chapter 11).

It is clear, too, that 'growth poles' serve less as diffusers of wealth and economic activity than focuses created in the urban hierarchy for the collection of new forms of social surplus to the benefit of the state-commercial ruling groups. To sum up, the dominant strategy of the past three decades, aimed at centralized modernization, has been one which has benefited a privileged minority, has consolidated existing power structures, has favoured the metropolitan centres at the expense of the regions and rural areas, and in part, through the persuasive influence of monopoly media systems, has broken down the obstacles to the further penetration of capitalist production, distribution and consumption.

We have referred to some of the debates in the critical writing on development over the past two decades, and have used some of their approaches in our studies of societies in Asia and Latin America. Part of our methodology has been to differentiate among the distinct historical phases of exploitation in Third World societies. By studying these phases in the evolution of capital and examining the relationship among capitalist and non-capitalist systems we have gained a greater depth and precision in our understanding of the factors constituting underdevelopment.

The impact of adventurers and merchants up to the eighteenth century, however harsh, was less profound and pervasive than that of nineteenth-century capitalism exporting its commodities, capital and ideological/administrative hegemony to the colonial territories. In turn, that phase was less intensive and disruptive than the direct expansion and economic/ideological control of twentieth-century corporations. With each step, the societies we have been comparing have been brought more thoroughly into the world system, whose trilateral centres, Europe-North America-Japan, compete constantly to extend their influence still further.[1]

We also believe that the major contradictions occur within rather than between nations, although the latter element enters the calculus.[2] Capital accumulation is taking place in all its forms by landowners, merchants, moneylenders and religious institutions, as well as the urban financial, commercial and industrial corporations, and, of course, transnational corporations. On the other hand, part of the surplus of the small-scale producers and urban lower circuit groups is being siphoned-off through the networks of primary accumulation, as shown in our Hong Kong and Ecuadorian examples, and by direct wage exploitation in the modern sector, as shown in our Malaysian study. The immediate beneficiaries are the local bourgeoisie. But the real gainers are those with whom the national state and local oligopolies maintain an uneasy and subaltern collaboration: the international corporate giants which form the spearhead of capitalism's expansion.

The dynamism of the present phase is impressive.[3] Feudal élites and merchants had always enjoyed a certain success at extracting surplus through coercion, ideological pressure or commercial monopoly, but the limits on their range of alternatives allowed the small producer a certain degree of independence. The extension of control by corporate capital directly into production, however, has dramatically changed the rules and expanded the limits of modern sector influence. It has also greatly extended the possibilities for capital accumulation through the creation of mass markets and extensive production of consumer durable products. This is especially true where corporations and collaborating governments have created artificial Free Trade Zones whose export-oriented industries benefit from the combination of high productivity and low wages for the workers there. With the internationalization of production capital, over the past two decades especially, surplus extraction from urban workers and rural wage-earners in commercialized agriculture has allowed for an impressive rise in the levels of capital accumulation in Third World societies.[4]

As we have shown in our studies, however, this does not preclude the continued existence of certain non-capitalist productive and commercial activity. In part, this is the result of a long struggle waged by a resilient small-scale producer class to retain its independence, but it is also a means by which the modern sector can keep wage costs down, and often avoid the social overhead costs of an industrial or unemployed workforce. Commercial agricultural employers can thus employ labour seasonally, leaving the workers to return to their peasant holdings which absorb the underemployed, the sick and the aged. Another advantage for employers is that unionization is less likely among the partially employed.

Much the same applies to urban workers in the modern sector of cities such as Cuenca, whose rural links provide the means to subsidize the low wages and also support for them during sickness and unemployment.[5] On the other hand, it is also true that the new (often young female) workers in the FTZs of countries such as Malaysia have the duty of helping to support their families in rural areas by sending home part of their wages. In this way, economic, social and cultural limits between cities and countryside, and of the capitalist and non-capitalist sectors, are becoming intertwined in an unequal relationship.

In summarizing the study, three main themes can be reiterated. All are connected with the internationalization of capital and the resultant concentration of economic activity, both structurally and geographically:

(a) The way in which concentration leads to convergence in the sphere of consumption in Third World societies and divergence in the sphere of production.

(b) The relative tendency of non-capitalist activity to dissolution or conservation by the capitalist mode of production.

(c) The specific function of the large city or central sub-system as a theatre of accumulation and diffuser of consumption patterns.

The first of these themes concerns the methods by which the processes of international capital accumulation have influenced the development of the urban systems, and, in particular, that of the primate cities on both continents. Roberts' (1978) investigation of convergence/divergence in Third World urban systems has been of value here. As secondary level decision-making bases within the international hierarchy of transnational operations (Hymer 1975), the cities of the upper system of Latin America and Asia are converging economically, socially and physically upon the models of the global metropolises.

As this occurs, they also begin to become distanced from the non-metropolitan regions; we can speak, therefore, of a convergence/divergence pattern of modernized production which accompanies the spread of capitalist accumulation. But there is one further element at play here which creates another type of convergence in Third World societies. Modernized consumption, the other face of capitalist production, has begun to permeate all levels of society with the goods and services of the capitalist sector. A form of diffusion is taking place, as conventional theory has argued it would, but it is a diffusion not of modern productive capacity so much as of new cultural patterns based on consumerism (Filgueira 1981). The end result is to accelerate further the process of dissolution in the lower circuits of country and town, often at the expense of the provision of basic needs such as staple foods.

The 'trickling down' of imitative consumerism promoted by a monopoly communications sector reinforces both dependence and inequality, strengthening the modern sector and giving greater leverage to the dominant metropolitan groups against the poor, the regions and the small producers, while diminishing the possibilities for more balanced, self-reliant and egalitarian development strategies.[6]

The approach adopted in this book can only be partially effective when the writers are not members of the social groups affected by the profound changes taking place: the small rural and village producer, the agricultural or factory worker, or the hawker in the lower circuit of the modernizing city. But we can be more certain of one question: that the immediate challenge to the present inequalities must arise *within* Third World societies. Alternative action has to begin by dealing with those internal conflicts of interest between the minority which appropriates and the majority which loses the surplus produced in the society.

And, in turn, this will require that those who have suffered most from the present strategy of modernized development attain the means by which they can break free from the structures of inequality and powerlessness which at present constrain them. The means for this lie in achieving control over their land and other means of production; participation in decision making; access to education; and the right to inform, as well as be informed, through democratic systems of communication.

We cannot finish this study without a brief comment on the possible outcomes of such developments. One consequence seems likely. Even as the principal cities have become the theatres for capital accumulation, some may also become the arenas for growing social dislocation. Latin America is now a predominantly urban continent, while the newly industrializing societies of Asia are concentrating their populations increasingly around their metropolitan centres. In one of our Asian case studies, urban concentration is almost total, while in the other, the FTZs and other forms of industrialization are accelerating the pace of urbanization. Even in Ecuador, one of the less urbanized societies of Latin America, nearly half the population is urban, with the larger part of that in only two cities.

The potential for alternative development strategies in these societies now seems to be tied to the dynamic of social relationships in the metropolitan centres (see Cardoso 1977, 21–39). This is not a hard and fast rule, of course, as experience in other countries has shown. The overthrow of the Argentine military in the early 1970s followed the *cordobazo* of a secondary city in 1968. More to the point, perhaps, the most profound social revolutions in Latin America have been brought about by insurrections of the dissident urban middle class leading rural guerrilla armies of peasants and workers.

From Cuba in 1959 to Nicaragua in 1979 and El Salvador today, this seems to have been the most drastic recipe for change.

It is not the only possibility, however, and it remains to be seen whether more gradual parliamentary experiments such as the Popular Unity interlude in Allende's Chile or the more recent return to traditional forms of democracy under the Radicals in Argentina will provide alternative models for change in the continent over the longer term.

Asian examples are rather more complex. In many societies the liaison (if somewhat uneasy at times) between the political élites, the armed forces, the new technocracy and local capitalists has worked to avoid intensifying class confrontations.[7] This has been greatly aided by the fact that many Asian countries have been able to deliver increased income to the population.

One further explanation for this relative lack of confrontation may be that many Asian market economies are at a stage of rapid economic expansion similar to that experienced by Argentina, Brazil and Uruguay in the period between 1880 and 1930. This would seem to suggest the likelihood of confrontation in which the urban working classes will play a major role in future social change.

In the short term even the World Bank recognizes the potential for social upheaval which is likely to arise from urban discontent, as Cheryl Payer has noted in her study of the institution:

> The daily contact between urban élites and low-income groups concentrated in urban areas where shelter is inadequate is potentially more explosive, politically and socially, than any problems of widely scattered rural settlements. (Payer 1982, 317)

Urban discontent is, of course, not easy to predict, and none of the trends we have discussed in this study would allow us to make categorical assertions concerning the pace and scope of social change. One thing is clear, however; the nature of capitalist expansion places cities in a central position in any predictive models of social and economic change. The decision makers of the national metropolises who are implementing modernization strategies have been concerned with global relationships for some time, but the horizons of artisans, hawkers, peasant and urban and rural wage-workers are now widening dramatically. Rural isolation has been broken down by networks of bus services, transistor radios and the influx of merchants, tourists and new commodities.

Urbanization itself has been stimulated by these changes and, at the same time, provides the environment for new levels of social activity. Political and social awareness is intensifying in many countries and evidence of more concerted forms of resistance to corporate penetration and authoritarian regimes whose legitimacy is being called into question is showing itself in different ways in societies as diverse as the Philippines, Argentina, Brazil, El Salvador and Guatemala. In others, this awareness is more muted, being countered in part by consumption-oriented state containment of political activity and the unformed nature of class development. The examples of Hong Kong, Taiwan and Singapore best illustrate this situation.

The path for each society is a unique one, and the outcomes will all be very different despite the homogenizing tendencies of capital. Nevertheless, through the turmoil and upheaval taking place, the interaction between capitalist modernization and the people it is affecting seems to be preparing the conditions for the creation of, in the words of Christopher Hill, 'a world turned upside down'.

Notes

1 *Third World urbanization: a changing scenario*

1 The shift through migration was seen to be occurring in many Third World societies, but the equation of urbanization and industrialization was considered to constitute *true* urbanization then. Castells (1977) criticizes the conventional concept of 'overurbanization' where 'the level of urbanization is higher than that which can "normally" be attained, given the level of industrialization'. This, he claims, is ethnocentric, applying 'the schema of economic growth of the advanced capitalist countries to other social forms in an entirely different conjuncture' (Castells 1977, 41).

The same criticism might also be applied to Hoselitz's (1953) concept of 'generative' and 'parasitical' cities where only capitalist sector growth is seen as 'generative'.

2 It is also true that the internationalization of capital was already under way as the United States, especially, expanded the range of direct overseas investment in productive activity throughout the Third World (see Roberts 1978, Castells 1977, Palloix 1977). Despite this, early dependency writers maintained that capitalism could not industrialize the periphery, a thesis challenged by Warren (1975). Of course, an earlier peripheral country, Japan, had bought technology and produced an internal industrial revolution which eventually led to patterns of urbanization remarkably similar to the west.

3 This is also discussed in the section of this chapter entitled New approaches to Third World urbanization.

4 We refer here not only to Singapore but also to the modernizing regimes of South Korea, Taiwan, Malaysia and Hong Kong, as well as Brazil, Mexico and Argentina, which are promoting export-oriented industrialization of their economies in co-operation with international firms and encouraged by agencies such as the IMF, the World Bank and consortia of private banks. The theoretical or indeed ideological justification for this open economy strategy has been provided largely by the monetarist school of economic thinking.

5 Former colleagues at the Department of Human Geography at Australian National University and one of the authors have attempted to do this in a study of food distribution and retailing in the urban areas of the New Hebrides. Although we make no claims for the thoroughness of our research methods, we managed to arrive at some estimates of the value of sales for various commodities through different types of outlets. This enabled us to arrive at some policy prescriptions designed to encourage the marketing and sale of indigenous foodstuffs (see McGee *et al.* 1980).

2 *Development theory and urbanization: rethinking the paradigm*

1 We are not, of course, suggesting that our paper caused this trend which arose out of a different set of concerns of international agencies and some orthodox development economists.

2 This contradiction continued to be a thorn in the flesh of attempts to modernize 'backward' economies. The 'Green Revolution' was supposedly one answer to the problems of traditional agriculture (see H. Cleaver 'The contradictions of the Green Revolution', 1973). The Alliance for Progress with its programme for continental agrarian change was another (also see J. Petras and R. Laporte 1973; D. Horowitz 1970).

The penetration of international capital has been the spearhead in the latest phase of agrarian modernization. Land tenure and production systems have been revolutionized by transnational corporate capital as part of the transformation of the traditional agricultural sectors of many Latin American countries in recent years (see R. Burbach and P. Flynn 1980; Lappé and Collins 1977; R. Ledogar 1976; S. George 1976, 1979).

3 See K. B. Griffin and A. R. Khan 1977.

4 This was further compounded in the early 1970s by the so-called energy crisis and the emergence of the 'eco-disaster' and 'growth for quality of life' schools.

5 Sceptically, one must wonder how international agencies could manage without this ability to develop crises. For some discussion of the meaning of the 'crisis mentality' to international agencies see McGee (1979a); Armstrong (1978); Ul Haq (1976).

6 The fact is that it is almost impossible to arrive at a precise definition of the Third World.

7 Bairoch (1974), questions the historical statistics usually offered as proof of declining terms of trade. He based his theoretical ideas on the work of Claude Raffestin and Mercedes Bresso (1974) and Andrew Pearse (1980).

8 Interesting collaborative evidence for Indonesia is produced by Collier *et al.* (1982).

9 This is a reformulation of the Baldwin (1956) argument that the potential for induced development in a society will be relatively greater in a family farm structure of production, where there is a more equal distribution of income, than in a plantation or debt-peonage system with a skewed pattern of income distribution. In a different way, T. Kemp (1972, 17), has emphasized that imperialism itself should not so much denote the relationship between dominant centres and dependent peripheries, but rather form the basis for analysing a particular stage in the international evolution of the capitalist mode of production.

10 One particular variation in recent years has been for the leaders of the armed forces to move overtly from their role as guardians of ruling civilian groups to establish themselves as a social class with their own economic interests in land, commerce, banking

and industry and with their own political formations. This has occured in Brazil and Indonesia among others, but a particularly striking case is that of Guatemala where army generals, US corporations and the State Department have combined to fill the political vacuum of the past three decades (Black 1983, 9ff.).

3 Cities: theatres of accumulation, centres of diffusion

1 Location theory and practice, with its emphasis on economies of scale in urbanization and localization economies, etc., is posited precisely on the existence of large urban entities with their labour, capital and consumer markets, technical, financial and professional services, links to government cultural facilities and so on. Nor have 'growth pole' theory/policy prescriptions altered the trend towards concentrated development. If anything, the strategy has exacerbated geographical, sectoral and social disparities (see Armstrong 1973, Gilbert 1980). On real estate investment see McGee (1985b).

2 This is not to argue that population growth is an unimportant factor in these processes. The fact that many Third World countries are cut off from the demographic safety valves of emigration clearly makes population growth a more significant element than it was in the period of rapid urbanization in western Europe.

3 The banking and financial systems, whether state or private, are at the very core of the upward, centralizing process of accumulation. They provide the mechanisms which channel capital away from the periphery to the centre, from peasant production to cash crop agriculture and to urban activities, from low-income groups to the wealthy, from unprofitable to 'growth' sectors. See Chapter 7 for an illustration of this process.

4 It is interesting that the two factors of low wages and lack of union strength are cited as important elements encouraging the rapid growth of industrialization even in regions of an advanced capitalist economy, such as the south of the USA (see Soja *et al.* 1983).

5 See Kwon Ping Ho (1979, 76–8) and T. Takeo (1978).

6 There are, of course, many regional and sectoral exceptions to the validity of this argument. For instance, it can be argued that the dissolution of petty commodity agriculture is occurring with extreme rapidity in some parts of Latin America (see Burbach and Flynn (1980) and Baird and McCaughan (1979) for the Mexican and Central American cases). On the other hand, the realities of social change in these societies may induce a realization that we are not dealing with clear alternatives. Rather, we note a creeping process in which independent producers are forced to take on wage employment to maintain their levels of family subsistence. Families thus straddle the divide between the modern sector and petty production activities, as our case studies on Southern Ecuador and Malaysia illustrate.

7 The argument, of course, is not against such policies but against the consequences of adopting such laws without making provision for local enterprises to adapt to them. There is also the argument that certain quality standards may be inappropriate to the needs of the dependent society.

8 If the Latin American Free Trade Agreement were allowed to operate freely, the dominance of this axis would be immensely increased (see Chapter 4 for further

comment). A similar situation would pertain with respect to Singapore in the ASEAN organization. Hymer (1975) deals with the hierarchical implications of this concentration/centralization process very effectively. See also Cohen (1981) and Friedmann and Wolf (1982).

4 *Accumulation and Latin American cities*

1 According to Tavares and Teixeira, there is thus established 'a convergent logic of accumulation in which public investment . . . becomes subordinate to investment by international enterprises, and both carry national private investment with them in their career' (1981, 96).

2 The consequences of such a strategy are summed up by L. C. Marinho who notes that the deepest tensions are experienced within industrializing societies, 'where the relative weight of the transnational corporations in the most dynamic sectors makes them particularly powerful agents. The necessary search for new dynamic activities aimed at satisfying the needs of broad sectors of the population could conflict with the interests of the world enterprises' branch plants, which are deeply committed to restrictive and "élitist" industrialization' (1981, 34).

3 R. Jenkins argues that 'The association between the TNCs and oligopolistic market structures lies at the heart of any analysis of their presence in the Mexican economy' (1979, 182).

4 However, as we have seen, the impact on consumption patterns may spread more widely over a range of social and regional groups as more and more factory-produced goods appear on local markets. We think that Santos' view must now be modified, although it was certainly more applicable to an earlier period.

5 The lower circuit may also play the role of introducing lower-income groups in the cities and in local markets to the goods produced by the capitalist sector. In this case, the local trader is acting as the intermediary in diffusing new patterns of consumption and goods to satisfy them, possibly at the expense of the local craftspeople.

6 Hymer continues

> The latter course was in fact chosen, and we now have the paradox that 500 million people can receive a live TV broadcast from the moon while there is still a shortage of telephones in many advanced countries, to say nothing of the fact that so many people suffer from inadequate food and lack of simple medical help.
>
> (Hymer 1975, 45)

7 As Roberts observes:

> It is not simply the self-interest of the dominant classes and of governments in Latin America . . . that perpetuates economic dependence; the type of social structure present in underdeveloped countries makes capital-intensive industrialization, oriented to sophisticated consumer goods, the easiest path to follow.
>
> (Roberts 1978, 79)

8 Santos concludes

> It is hardly surprising that the large, developed country's enterprises locate parts of their production-process in depressed regions, especially when the state, through regional development programmes and other incentives, creates such a favourable regional environment for foreign investment. Modern externally oriented industries can be located without consideration of the potential or actual domestic market, simply because their market lies elsewhere. (Santos 1979, 172)

9 Not in all, however. The open market, monetarist strategy of the Chilean military government has decimated local manufacturing industries without managing to attract foreign corporate capital to modernize the national economy. Similarly, the modernization drive in societies such as Nicaragua and El Salvador in the 1960s and 1970s exacerbated social and political tensions to the extreme of civil war, a counterproductive result for capital which seeks stability for its operations.

5 *The urbanization process, accumulation and industrialization in the market economies of Asia*

1 While it is a contentious question because of the significant participation of the state in market economies, we have taken all the major countries in Asia (with the exception of Japan) where private ownership still prevails and defined them as market economies. The following countries have been included in the analysis: Hong Kong, Singapore, Sri Lanka, Bangladesh, Burma, India, Indonesia, Pakistan, South Korea, Malaysia, Philippines, Taiwan and Thailand. The investigation of the Japanese experience of urbanization is very important but, in our view, is too large a topic to tackle in this study.

2 The problems of accurate definition of the labour force participation in various industrial sectors is very complicated in the Asian context where many people engage in more than one activity and household labour inputs are not always recorded adequately. See Hauser (1971, 1972); McGee (1979b, 1981a); Breman (1976, 1978).

3 The most comprehensive discussions of Asian development have emerged in a series of studies by Lo and Salih (1978, 1979) which develop a broad model of the types of Asian economies taking into account these various factors.

4 A theme developed at length with respect to developed countries in Castells (1977), Harloe (1977), Tabb and Sawers (1978), Dear and Scott (1981).

5 This section paraphrases Harvey (1981), pp. 94–101.

6 This was not true of the city-state of Hong Kong.

7 These figures are cited in current US dollars (see Keesing 1979, Table 6).

8 See UNCTAD (1978), Table 2.

6 *Ecuador: accumulation and the producers*

1 As Bromley states:

> Ecuador's cultural dependence has encouraged reliance on foreign investment and expertise, the unnecessary import of foreign-produced goods and technologies,

> conspicuous consumption by local élites, and the emigration of local technicians. It has acted as a disincentive to domestic industrialization and to domestic entrepreneurship, and has encouraged Ecuador's successive governments to copy development styles and strategies from North America and Europe rather than try to devise strategies more suitable to local conditions. (Bromley 1977, 6)

2 The history of attempts to change land tenure by public intervention is a sad one. The first agrarian reforms date from the 1960s when Ecuador was stirred by Alliance for Progress to change the inequitable distribution of land. The first reform law was promulgated in 1964 by a moderate military government, then the issue was virtually ignored until 1970 when another law was passed – with essentially the same results. For a short review of this period see Charles S. Blankstein and Clarence Zuvekas (1973) who are mild in their criticism. But it is evident that any real possibility for altering the agrarian structure has foundered on the opposition of landowners, who reduced the reforms to minor redistribution of the poorest lands, some colonization, and a pilot project covering 11,000 hectares in the Guayas river basin. If anything, the laws of 1964 and 1970 probably had negative effects on the welfare of the *campesinos*. Under such circumstances, the authors' summarizing comment that: 'Agrarian reform efforts in Ecuador to date have been very disappointing' (p. 94) seems, at the very least, restrained.

3 The information for this section and the studies of Suscal, El Tambo, Deleg, Gima and other localities are based on interviews and discussions with teachers, farmers, development agency officials, artisans, peasants, shopkeepers and other business people.

4 See also E. Feder (1976a), in which he argues similarly that low interest rates subsidize the large-scale producers in agriculture and exacerbate existing disparities between the social classes.

5 Of course, this is not to say that such absentee landowners in the city of Cuenca live exclusively from such rurally derived surplus. As we shall see later, many have now diversified their sources of income through investments in urban commercial, financial and real estate activity. This does not negate the fact, however, that the process of primary accumulation in the agricultural sector continues at the expense of the peasantry.

6 Charles Bettelheim comments on this type of unequal exchange under conditions of peripheral capitalist evolution:

> in countries where there are a large number of producers who produce for the market only to a minor extent, the prices of the product placed on the market under these conditions have an extremely complex relation to their cost in money or in labour, for these costs do not directly play the role of regulator. It is thus possible for prices to fall, for an indefinite period below these costs of production. . . . The peasants who are in debt are *obliged to produce and sell*, even at prices that merely give them the means to survive and to pay off, at least in part, the *burden of debt* that weighs upon them. This is often the situation of the peasants who own very small pieces (parcels) of land. The role played by usurer's capital, and later by banking capital, as the dominant element in the reproduction process, is particularly obvious here. (Bettelheim 1972, 298–9)

7 For a further discussion of this last migration phenomenon see Breman (1976, 1978).

8 Samir Amin argues that a real change from the dependent forms of development in peripheral capitalist societies would turn this whole exploitative process upside down. 'First', he writes, 'we must go into reverse gear. Up to now industry in the Third World has been parasitic, in the sense that it has based its accumulation on extortions from the rural world in real terms (it gets its labor from the rural-urban migration) and financial terms (heavy taxation, internal terms of trade unfavorable to the peasants, etc.) with no counterpart provided in return to sustain a take-off of agriculture.' To do this requires two primary steps: first, the linking of a 'modern' industrial sector with small-scale rural industries 'which can directly mobilize the latent forces of progress' and second, rural collectivization and *not* 'private' agriculture 'even if remodeled by a radical land reform' (Amin 1977, 16–17).

9 Personal communication from a factory social worker and development adviser in Cuenca.

10 The Cuenca daily, *El Mercurio* (9 June 1976) comments aggrievedly on this new trend in the city's modernizing experience.

7 *Ecuador: capital and the accumulators*

1 On linguistic dependence, Louis Jean Calvet writes:

> any nominal liberation which is not accompanied by an overturning of the linguistic superstructure is not a liberation of the social class which spoke and continues to speak the dominating language. . . . It is only after they have gotten rid of their comprador bourgeoisie, after having broken the capitalist state, that the neo-colonial peoples, liberated colonially *and* socially, will find simultaneously the mastery of their destiny and of their language (or, in a wider and more correct fashion: the mastery of their civilization). (Calvet 1974, 137)

For the speakers of Quechua or the indigenous languages of the Oriente, the separation from the languages of power and influence is doubled. Not only are they strangers to Spanish, the language of colonialism; they are totally excluded from the language of neo-colonialism, English.

2 Guillermo Navarro (1976, 49–50) comments on the links between the Banco del Azuay as lender and the firm, Importadora Mercantil Mirasol, as borrower, the personal nexus between the two being Senora Francisca Vega de Arízaga, with shares in both.

3 Navarro argues that it is the banking sector, above all, which provides the link between foreign investment and local capital; 'it can be concluded that it is the banks which dynamize the process of "alliance between national and foreign capital"' (1976, 47).

4 The full picture of the Arízaga participation in the Agency's Ecuadorian and Latin American operations is given in Philip Agee (1975).

5 The significance of this lies in the relationship between the capital of the Cuenca merchants and the industrial capital of the largely foreign capitalist sector. The former can exist in the sphere of trade or circulation without changing the mode of producing the goods while the latter alters the modes and relations of production in society.

In the case of Cuenca's merchants, the least painful and most immediately profitable way of making the change to new activities, especially at the national scale, has been to associate with foreign capital. Industrialization has, therefore, largely been characterized by dependence upon foreign technology, management and decision making.

6 See Navarro (1976, 44–5). The reasons for such a concentration stem from external economies, access to finance and market and the exercise of political influence in the centres of power. The value of these to the corporation is undoubted; the disadvantages for outlying areas equally so.

7 Of the five companies in which the Cordero Crespo family has majority control, one is in the province of Pichincha (Quito), two in Guayas (Guayaquil) and two in Azuay (Cuenca).

8 See the writings of Fernando Velasco in the journal *Economia*, 1973a, 1973b especially, for analyses of Ecuador's history of agricultural export dependence.

9 In the Third World, modifications are taking place as a result of the penetration of direct monopoly capital: according to Poulantzas (1974), capitalism

> dominates these formations not simply from the 'exterior' and by the means of the reproduction of the *relation* of dependence *but establishes its direct domination within them. The mode of production of the metropoles is reproduced in a specific form in the very interior of the dominated and dependent formations*. This presents no obstacle to the ability of the effects of conservation, [however] . . . to override those of dissolution. (Poulantzas 1974, 150–1)

The applicability of both Petras' and Poulantzas' analysis to Ecuador through the 1970s is evident.

10 Daniel Defoe, *Jure Divino*, in Christopher Hill (1972, 381).

11 Much of the following description is derived from interviews in the region.

12 The fiestas throughout the year are already numerous, but the Church still adds to them from time to time. It is interesting, therefore, to see a commentary arguing against further expansion in the conservative *El Mercurio* of 31 May 1976.

13 Philip Agee writes of an instance in 1962 when villagers in a community close to Cuenca attacked and killed members of a medical team from the Andean Mission, an organization supported by the United Nations. The local priest refused to intervene, having himself spread the story that the team was communist (Agee 1975, 253–4).

14 Personal communication, June 1976. It is not the case, though, that the Indian communities invariably accept the guidance and orders of the priest. Ecuadorian history is full of examples of Indian revolts against oppression, especially when it concerned rights to the ownership of land. The theme has been discussed in detail by Pio Jaramillo (1954) and by Luis Monsalve (1943).

'The Indian', Monsalve says, 'is the land. The coupling of the being and the thing is here absolute and total. And this coupling explains the acceptance by the Indian of his slavery.'

But, equally, the bond with the land explains the occasional Indian uprisings when the threat to the land became clear. These rebellions and their repression are dealt with by Jaime Galarza (1975), Albornoz (1976) and Jorge Icaza in his powerful novel, *Huasipungo* (1953).

15 Such work is undertaken in the face of open hostility by public authorities. See Hayes Keeler, Rodrigo Mera and Roberto Cruz (1973).

Note that such agricultural co-operatives are often undermined as a result of the encouragement by government officials of individualistic practices among their members and that attempts to market co-operative products are often harassed by municipal authorities.

16 For a detailed discussion of these operations see Armstrong (1978).

17 Most of the material on CREA is derived from CREA (1976) and the report of the United Nations Regional Development advisory mission (UNOTC 1974).

18 The statement by the then-new Director of CREA was delivered at a meeting at CREA headquarters, Cuenca, 27 May 1976.

8 *Conservation-persistence in the two-circuit system of Hong Kong: a case study of hawkers*

1 See Chairman, Urban Council and Director, Urban Services Department, *Annual Departmental Report, 1967–8*, 30.

2 See McGee (1970). Note that McGee estimated a ratio of 3.2 hawkers per 100 of population for Hong Kong hawkers based on the prevailing estimates of hawkers and population. In the light of the figures now available that was inaccurate.

3 In the case of pedlars this means the same location or grouping of hawkers, for they do not always move to the identical pitch every day. It is generally agreed among hawkers that some pitches are more lucrative than others, particularly those at the entrance to agglomerations, but pedlars tend to share out these more lucrative pitches over a period of time by mutual agreement.

4 Interviews and correspondence with government officials.

5 See Hong Kong Government Census and Statistics Department (1972), Table 12.

6 Hong Kong Urban Council Proceedings, 2 June 1964, 99.

9 *Women workers or working women? A case study of female workers in Malaysia*

1 See Lim (1982). Salaff (1981) has a fascinating discussion of some of these themes in her case history of 'Rainbow'.

2 The authors recognize that there are much broader conceptual models of proletarianization, particularly those relating to the development of a proletarian class and consciousness. See McGee (1985a) for a review of these theories.

3 This term is used in the broadest sense to include 'free trade' territories, such as Hong Kong, as well as specific areas set aside for the manufacturing and assembly of exports where special conditions operate for tariffs and labour. These areas are also called Free Trade Zones and Free Export Zones, according to individual nations' definitions.
4 See the bibliographic review of Lim (1980) and O'Brien (1981).
5 See Safa (1980). O'Brien (1982) describes the process in Western Europe.
6 The Singapore case is analysed in Salaff and Wong (1982), Wong (1979).
7 HAWA is an acronym for 'haul wanita' meaning strength or power for women. The results of the HAWA project are reported in Lim (1982).
8 These following statistics are extracted from the *Annual Report 1980* published by the Penang Development Corporation. Hardly surprisingly they indicate success. We await the access to the 1980 Census to provide additional information.
9 This is the type of strategy employed by many of the Hong Kong families described by Salaff (1981). Cultural values will also be important to investigate in looking at the family based strategies.
10 Indeed, these questions are the basis of current research being conducted by one of the authors, McGee, and a team of researchers from University Sains, Penang, led by Dr Kamal Salih.
11 See Grossman (1979) for a summary of these arguments.

10 Conclusion

1 We might note that the rivalry among the three major centres in the world system caused sufficient concern to bring leading members of the political, business, trade union and academic spheres together in the 1970s to form a 'community of developed nations' which might be able 'to co-ordinate international, political and economic affairs' (Novak 1977). For an inside view of the aims of the Trilateral Commission see the report of M. J. Crozier, S. P. Huntingdon and J. Watanuki (1975), and for a later outside assessment, H. Sklar (1980).
2 Wolfe (1976) also criticizes the concept of central and peripheral nations and the division between 'rich' and 'poor' countries. The latter distinction has some value in classifying societies internationally, but is erroneous in arousing

> the supposition that each group of countries is homogenous in essential characteristics, with common interests and problems. In fact, both groups are highly heterogeneous and increasingly so in their power structures, ideologies, basic resources, population structures and their role in the international order.
>
> (Wolfe 1976, 14)

3 This statement may appear strange at a time of contraction in the international capitalist system but it is our view that this contraction masks a continuing expansion on the part of transnational corporations, especially in the new growth sectors such as electronics and information systems.
4 For a case study of the capacity of agribusiness to raise profit levels by depressing wages while raising productivity on their Mexican holdings see Baird and McCaughan (1979), especially the chapter 'Runaway crops'. Their equivalent study of the cheap labour policies of the Mexican FTZs on the frontier with the United States, where

the *maquiladora* workers receive salaries one-fifth the level of their North American counterparts, appears in the chapter 'Runaway shops'.

5 See Claude Meillasoux (1972), who writes that 'the preservation of the relations with the village and the familial community is an absolute requirement for the wage earners, and so is the maintenance of the traditional mode of production as the only one capable of ensuring survival' (p. 98).

6 Juan Somavía (1981) writes:

> There is today a minority social monopoly over communication means in Latin America. The property and control of the principal industrial instruments of information are basically in the hands of one social sector: the powerful private economic and financial interests which exert major influence in national affairs. Through direct ownership and a variety of other methods, they can orient the content and general direction of mass media and have the power to influence significantly the total information agenda of the societies in the region.
>
> (Somavía 1981, 17)

7 This is not to deny the existence of active guerrilla movements in countries such as the Philippines, but they are not nearly as pervasive as in Latin America.

Bibliography

General and theoretical

Adam, G. (1975) 'Multinational corporations and worldwide sourcing', in H. Radice *International Firms and Modern Imperialism: Selected Readings*, Harmondsworth, Penguin, 89–103.

Althusser, L., and Balibar, E. (1970) *Reading Capital*, London, New Left Books.

Amin, S. (1974) *Accumulation on a World Scale: A Critique of the Theory of Underdevelopment*, 2 vols. New York, Monthly Review Press.

—— (1976) *Unequal Development*, New York, Monthly Review Press.

—— (1977) 'Self-reliance and the new international order', *Monthly Review*, 29, 3, 1–21.

Anderson, P. *Lineages of the Absolutist State*, London, New Left Books.

—— (1980) *Arguments within English Marxism*, London, Verso (New Left Books).

Armstrong, W. R. (1980) 'Land, class, colonialism: the origins of dominion capitalism', in W. E. Willmont (ed.) *New Zealand and the World: Essays in Honour of Wolfgang Rosenberg*, Christchurch, University of Canterbury, 28–44.

—— (1981a) 'The Bretton Woods time lag: monetarism in Argentina, Chile and Australia thirty years on', *USG Newsletter*, McGill University (J. Bradbury, Department of Geography), April, 58–69.

—— (1981b) 'The industrialization of staples producers: Australia, Canada and Argentina in the nineteenth century', *Occasional Paper 8*, Sydney, Planning Research Centre, Sydney University.

—— and Bradbury, J. H. (1983) 'Industrialization and class structure in Australia, Canada and Argentina, 1870–1980', in E. L. Wheelwright and K. Buckley (eds) *The Political Economy of Australian Capitalism*, vol. 5, Sydney, ANZ Press.

—— and McGee, T. G. (1968) 'Revolutionary change and the Third World city: a theory of urban involution', *Civilisations*, 18, 353–77.

Bairoch, P. (1974) *Révolution industrielle et sous-dévéloppement*, Paris, Mouton.

Baldwin, R. E. (1956) 'Patterns of development in newly settled regions', *Manchester School of Economic and Social Studies*, vol. XXIV, May, 161–79.
Banaji, J. (1977) 'Modes of production in a materialist conception of history', *Capital and Class*, 3, Autumn 1–44.
Baran, P. (1957) *The Political Economy of Growth*, New York, Prometheus.
Barratt-Brown, M. (1974) *The Economics of Imperialism*, Harmondsworth, Penguin.
Bechhofer, F. and Elliot, B. (1968) 'An approach to a study of small shopkeepers and the class structures', *Archives europées de sociologie*, tome IX, 180–202.
Belshaw, C. W. (1965) *Traditional Exchange and Modern Markets*, Englewood Cliffs, NJ, Prentice-Hall.
Bergeson, Albert (1982) 'The emerging science of the world system', *International Social Science Journal*, xxxiv, 1, 23–36.
Bettelheim, C. (1972) 'Theoretical comments', in A. Emmanuel, *Unequal Exchange*, New York, Monthly Review Press.
Bienefeld, M. A. (1975) 'The informal sector and peripheral captalism', *Bulletin, Institute of Development Studies*, 6, 3, 53–73.
Billerbeck, K. and Yasugi, Y. (1979) *Private Direct Foreign Investment in Developing Countries*, World Bank Staff Working Paper 348, Washington, DC, World Bank.
Boeke, J. H. (1953) *Economics and Economic Policy of Dual Societies*, New York, Institute of Pacific Relations.
Booth, D. (1975) 'André Gunder Frank: an introduction and appreciation', in I. Oxaal and D. Booth (eds) *Beyond the Sociology of Development*, London and Boston, Routledge & Kegan Paul, 50–85.
Bowles, S. (1978) 'The Trilateral Commission: have capitalism and democracy come to a parting of the ways?', in Union of Radical Political Economists, *US Capitalism in Crisis*, New York, URPE, 261–75.
Bradby, B. (1975) 'The destruction of natural economy', *Economy and Society*, 4, 2, 127–61.
Braverman, H. (1974) *Labor and Monopoly Capital: The Degradation of Work in the Twentieth Century*, New York, Monthly Review Press.
Breman, J. (1976) 'A dualistic labour system? A critique of the informal sector concept', *Economic and Political Weekly* XI (48): 1870–5; XI (49): 1905–08; XI (50): 1939–44.
—— (1978) 'Seasonal migration and cooperative capitalism', paper for international seminar on the cross-cultural study circulation, Honolulu, East-West Population Institute, East-West Center.
Brenner, R. (1977) 'The origins of capitalist development: a critique of Neo-Smithian Marxism', *New Left Review*, 104, 25–93.
Bromley, R. J. and Gerry, C. (eds) (1979) *Casual Work and Poverty in Third World Cities*, London, Methuen.
Brookfield, H. (1975) *Interdependent Development*, London, Methuen.
Brown, L. R. (1977) 'The urban prospect: re-examining the basic assumptions', *Population Development Review*, 2, 267–77.
Browning, H. L. (1958) 'Recent trends in Latin American urbanization', *Annals of the American Academy of Political and Social Science*, March, 111–20.
—— and Roberts, B. R. (1980) 'Urbanization, sectoral transformation, and the utilization of labor in Latin America', *Comparative Urban Research*, VIII, 1, 86–103.
Burn, R. (1978) 'The changing world of foreign banks', *The Banker*, April, 45–54.

Cabral, A. (1973) *Return to the Source: Selected Speeches of Amilcar Cabral*, New York, Monthly Review Press and Africa Information Service.

Calvet, L. J. (1974) *Linguistique et colonialisme*, Paris, Payot.

Cardoso, F. H. (1975) 'The city and politics', in J. E. Hardoy (ed.) *Urbanization in Latin America: Approaches and Issues*, New York, Anchor Books, 157–90.

—— (1977) 'Towards another development' in M. Nerfin (ed.) *Another Development: Approaches and Strategies*, Uppsala, Dag Hammarskjold Foundation, 21–39.

—— and Faletto, E. (1969–78) *Dependencia y Desarrollo en América Latina*, Mexico City, Siglo XXI Editores, 1969; (*Dependency and Development in Latin America*, Berkeley and Los Angeles, University of California Press, 1978).

Castells, M. (ed.) (1973) *Imperialismo y Urbanización en América Latina*, Barcelona, Editorial Gustavo Gili.

Castells, M. (1977) *The Urban Question: A Marxist Approach*, London, Edward Arnold.

Chayanov, A. V. (1966) *The Theory of the Peasant Economy*, trans. and ed. by D. Thorner, R. E. F. Smith and B. Kerblay. Homewood, Ill., published for the American Economic Association by Richard D. Irwin Inc.

Clairmonte, F. F. and Cavanagh, J. (1982) 'Corporate power in selected food commodities', *Raw Materials Report*, 1, 3, 21–40.

Clark, C. (1951) *The Conditions of Economic Progress*, London, Macmillan.

Cleaver, H. (1973) 'The contradictions of the Green Revolution', in C. K. Wilber (ed.) *The Political Economy of Development and Underdevelopment*, New York, Random House, 187–97.

Cohen, R. B. (1981) 'The new international division of labor, multinational corporations and urban hierarchy', in M. Dear and A. J. Scott (eds) *Urbanization and Urban Planning in Capitalist Society*, London, Methuen, 287–315.

Connell, R. W. and Irving, T. H. (1980) *Class Structure in Australian History: Documents, Narrative and Argument*, Melbourne, Longman Cheshire.

Coraggio, J. L. (1972) 'Hacia una revisión de la teoría de los polos de desarrollo', *Revista Latino-Americana de Estudios Urbanos y Regionales (Eure)*, Santiago, 2.

Crozier, M. J., Huntingdon, S. P. and Watanuke, J. (1975) *The Crisis of Democracy: Report on the Governability of Democracies to the Trilateral Commission*, New York, New York University Press.

Dear, M. J. and Scott, A. J. (eds) (1981) *Urbanization and Urban Planning in Capitalist Society*, London, Methuen.

De Marco, S. and Sechler, S. (1975) *The Fields Have Turned Brown – Four Essays on World Hunger*, Washington, DC, The Agribusiness Accountability Project.

De Mattos, C. A. (1982) 'The limits of the possible in regional planning', *Cepal Review*, 13, December, 65–86.

Dillinger, W. (1979) *A National Urban Data File For 114 Countries*, Urban and Regional Report 79–5, Washington, DC, World Bank.

Dobb, M. (1963) *Studies in the Development of Capitalism*, London, Routledge & Kegan Paul.

Dorfman, A. and Mattelart, A. (1975) *How to Read Donald Duck: Imperialist Ideology in the Disney Comic*, New York, I. G. Editions Inc.

Dos Santos, T. (1973a) *Dependencia y Cambio Social*, Buenos Aires, Amorrortu Editores.

Dos Santos, T. (1973b) 'The crisis of development theory and the problem of dependence in Latin America', in H. Bernstein (ed.) *Underdevelopment and Development in the Third World Today*, Harmondsworth, Penguin, 57–80.

Dunleavy, P. (1980) *Urban Political Analysis: The Politics of Collective Consumption*, London, Macmillan.

Ehrensaft, P. and Armstrong, W. R. (1978) 'Dominion capitalism: a first statement', *The Australian and New Zealand Journal of Sociology*, 14, 3.

—— and Armstrong, W. R. (1981) 'The formation of dominion capitalism: economic truncation and class structure', in A. Moscovitch and G. Drover (eds) *Inequality: Essays on the Political Economy of Social Welfare*, Toronto, University of Toronto Press, 99–155.

Emmanuel, A. (1972) *Unequal Exchange: A Study of the Imperialism of Trade*, New York and London, Monthly Review Press.

Ewen, S. (1976) *Captains of Consciousness: Advertising and the Social Roots of Consumer Culture*, New York, McGraw Hill.

Feder, E. (1976a) 'How agribusiness operates in underdeveloped agricultures: Harvard Business School myths and reality', *Development and Change*, 7, 413–43.

—— (1976b) 'The new Third World program for the self-liquidation of the Third World peasantry', *Journal of Peasant Studies*, 3, 3, 343–54.

Feldman, A. and Moore, W. (1962) 'Industrialization and industrialism: convergence and differentiation', *Transactions of the Fifth World Congress of Sociology*, 2, International Sociological Society.

Filgueira, C. (1981) 'Consumption in the new Latin American models', *Cepal Review*, 15, December 71–110.

Finger, J. M. (1975) 'Tariff provisions for offshore assembly and the exports of developing countries', *Economic Journal*, 85, 365–71.

Food and Agriculture Organization (1980) *FAO Production Yearbook, 1980*, New York, United Nations.

Frank, André Gunder (1980) *Crisis in the World Economy*, New York, Holmes & Meier.

—— (1981) *Crisis in the Third World*, New York, Holmes & Meier.

Franklin, S. H. (1969) *The European Peasantry: The Last Phase*, London, Methuen.

Fraser, W. H. (1981) *The Coming of the Mass Market, 1850–1914*, Hamden, Conn., Archon Books.

—— and Sullivan, F. (1974) 'The absorption of labor in the urban economy: the case of developing countries', *Economic Development and Cultural Change*, 22, 385–413.

Friedmann, J. and Wolf, G. (1982) 'World city formation: an agenda for research and action', *International Journal of Urban and Regional Research*, 6, 3, 309–44.

Fröbel, F., Heinrichs, J. and Kreye, O. (1977) 'The tendency towards a new international division of labor: the utilization of a world-wide labor force for manufacturing oriented to the world market', *Review*, 1, 1, 73–88.

——, ——, —— (1980) *The New International Division of Labour*, Cambridge/Paris, Cambridge University Press/Editions de la Maison des Sciences de l'Homme.

Geertz, C. (1963) *Peddlers and Princes*, Chicago, University of Chicago Press.

George, S. (1976) *How the Other Half Dies: The Real Reasons for World Hunger*, Harmondsworth, Penguin.

—— (1979) *Feeding the Few: Corporate Control of Food*, Washington, DC, Institute for Policy Studies.

Gerry, C. (1974) *Petty Producers and the Urban Economy: A Case Study of Dakar*, Research Working Paper, 8, World Employment Programme, Geneva, International Labour Office.

Gilbert, A. and Gugler, J. (1982) *Cities, Poverty and Development: Urbanization in the Third World*, London, Oxford University Press.

Girvan, N. (1976) *Corporate Imperialism: Conflict and Expropriation*, New York, Monthly Review Press.

Goldsmith, W. W. (1977) 'The war on development', *Monthly Review*, 28, 10.

Goodman, D. and Redcliffe, M. (1981) *From Peasant to Proletarian: Capitalist Development and Agrarian Transitions*, Oxford, Basil Blackwell.

Gorz, A. (1975) 'Colonialism at home and abroad', *Our Generation* (Montreal), 9, 2.

Griffin, K. B. (1976) *Land Concentration and Rural Poverty*, London, Macmillan.

—— and Khan, A. (1977) 'Rural poverty in developing countries', paper given at the International Economic Association Conference, Tokyo, August-September.

Hall, S. (1981) 'In defence of theory', in R. Samuel (ed.) *People's History and Socialist Theory*, London, Routledge & Kegan Paul, 378–85.

Harloe, M. (ed.) (1977) *Captive Cities*, London, John Wiley.

Hart, K. (1973) 'Informal income opportunities and urban employment in Ghana', *Journal of Modern African Studies*, II, 1, 61–89.

Harvey, D. (1973) *Social Justice and the City*, London, Edward Arnold.

—— (1981) 'The urban process under capitalism: a framework for analysis', in M. Dear and A. J. Scott (eds) *Urbanization and Urban Planning in Capitalist Society*, London and New York, Methuen, 93–121.

Hauser, P. M. (1971) 'A new approach to the measurement of the work force in developing areas', unpublished paper (limited circulation).

—— and Gardner, R. W. (1980) 'Urbanization, urban growth and intermediate cities: trends and prospects', paper prepared for East-West Population Institute Workshop on Intermediate Cities, Honolulu, July.

—— and Schnore, L. F. (eds) (1965) *The Study of Urbanization*, New York, Wiley.

Helleiner, G. K. (1973a) 'Manufactured Exports from the LDC's and Multinational Firms', *Economic Journal*, 83.

—— (1973b) 'Manufacturing for export, multinational firms and economic development', *World Development*, 1.

—— (1975) 'The role of multinational corporations in the less developed countries' trade and technology', *World Development*, 3.

—— (1979) 'Intra-firm trade and the developing countries: an assessment of the data', *Journal of Development Economics*, 6, 3.

Hill, C. (1972) *The World Turned Upside Down: Radical Ideas During the English Revolution*, Harmondsworth, Penguin.

Hilton, R. (ed.) (1976–8) *The Transition from Feudalism to Capitalism*, London, Verso (New Left Books).

Hoselitz, B. F. (1953) 'The role of cities in the economic growth of underdeveloped countries', *Journal of Political Economy*, 61, 195–203.

Hymer, S. (1975) 'The multinational corporation and the law of uneven development', in H. Radice (ed.) *International Firms and Modern Imperialism: Selected Readings*, Harmondsworth, Penguin, 37–62.

International Labour Office (1972) *Employment, Incomes and Equality: A Strategy for Increasing Productive Employment in Kenya*, Geneva, International Labour Office.

Jacoby, E. H. (1970) 'The coming backlash of semi-urbanization', *Ceres*, 18, 48–51.

Johnson, D. L. (1981) 'Economism and determinism in dependency theory', *Latin American Perpsectives*, VII, 3 & 4, 108–17.

Johnson, R. (1981) 'Against absolutism', in R. Samuel (ed.) *People's History and Socialist Theory*, London, Routledge & Kegan Paul, 386–96.

Katzin, M. F. (1959). 'The Jamaican country higgler', *Social and Economic Studies*, 8, 421–40.

Kay, G. (1975) *Development and Underdevelopment: A Marxist Analysis*, London, Macmillan.

Keesing, D. B. (1979) 'World trade and output of manufacturers: structural trends and developing countries' exports', World Bank Staff Working Paper, 316, Washington, DC, World Bank.

Kemp, T. (1972) 'The Marxist theory of imperialism', in R. Owen and B. Sutcliffe (eds), *Studies in the Theory of Imperialism*, London, Longman, 15–34.

Kondratieff, N. D. (1935) 'The long wave of economic life', *The Review of Economic Statistics*, 17, 105–15.

Lappé, F. M. and Collins, J. (1977) *Food First: Beyond the Myth of Scarcity*, Boston, Houghton Mifflin.

Le Brun, D. and Gerry, C. (1973) 'A theoretical prelude to class analysis of petty producers in Senegal', unpublished paper, Geneva, International Labour Office.

Lewis, W. A. (1958) 'Economic development with unlimited supplies of labour', in A. N. Agarwala and S. P. Singh (eds) *The Economics of Underdevelopment*, Oxford, Oxford University Press, 400–49.

Linsky, A. S. (1965) 'Some generalizations concerning primate cities', *Annals of the Association of American Geographers*, 55, 506–13.

Lipton, M. (1977) *Why Poor People Stay Poor: Urban Bias in World Development*, Canberra, Australian National University Press.

Lloyd, P. (1982) *A Third World Proletariat*, London, George Allen & Unwin.

Logan, M. and Salih, K. (1982) 'Changes in the world economic system and their implications for urban development policy', paper presented to Conference on Urbanization and National Development, East-West Population Institute, Honolulu.

McGee, T. G. (1971c) *The Urbanization Process in the Third World*, London, Bell.

—— (1973) 'Peasants in the cities. A paradox, a paradox, a most ingenious paradox', *Human Organization*, 32, 2, 135–42.

—— (1976) 'The persistence of the proto-proletariat. Occupational structures and planning of the future of Third World cities', *Progress in Geography*, 9, 1–38.

McGreevey, W. P. (1971) 'A statistical analysis of primacy and log normality in the size distribution of Latin American cities, 1750–1960', in R. M. Morse (ed.) *The Urban Development of Latin America, 1750–1920*, Stanford Calif. Stanford University, Centre for Latin American Studies.

Mandel, E. (1976) 'Introduction', in K. Marx (1976) *Capital*, vol. I, Harmondsworth, Penguin.

—— (1980) *The Long Waves of Capitalist Development*, Cambridge, Cambridge University Press.

Mangin, W. (ed.) (1970) *Peasants in Cities, Readings in the Anthropology of Urbanization*, Boston, Houghton Mifflin.

Meillasoux, C. (1972) 'From reproduction to production: a Marxist approach to economic anthropology', *Economy and Society*, 1, 1.

Mendleson, M. S. (1980) *Money on the Move*, New York, McGraw Hill.

Mintz, S. W. (1961) 'Pratik: Haitian personal economic relations', in *Proceedings of the Annual Spring Meeting of the American Ethnological Society*, Seattle, University of Washington Press.

—— (1971) 'Men, women and trade', *Comparative Studies in Society and History*, 13, 247–69.

Mitra, J. D. (1979) 'The capital goods sector in LDCs: a case for state intervention', World Bank Staff Working Paper, 343, Washington, DC, World Bank.

Mosley, P. (1977) *Implicit Models and Policy Recommendations: Reflections on the Employment Policy in Kenya*, paper prepared for Institute of British Geographers' Symposium on the Urban Informal Sector in the Third World, School of Oriental and African Studies, University of London.

Munck, R. (1981) 'Imperialism and dependency: recent debates and old dead-ends', *Latin American Perspectives*, VIII, 3 & 4, 162–79.

Nayyar, D. (1978) 'Transnational corporations and manufactured exports from poor countries', *The Economic Journal*, 88, 59–84.

Nelson, J. M. (1976) *Temporary versus Permanent Citywards Migration: Causes and Consequences*, Cambridge, Mass., MIT Press.

—— (1979) *Access to Power, Politics and the Urban Poor in Developing Nations*, Princeton, NJ, Princeton University Press.

Novak, J. (1977) 'The trilateral connection', *Atlantic Monthly*, July, 57–9.

Nurkse, R. (1962) *Patterns of Trade and Development*, Oxford, Basil Blackwell.

O'Brien, P. J. (1975) 'A critique of Latin American theories of dependency', in I. Oxaal, T. Barnett and D. Booth (eds) *Beyond the Sociology of Development*, London and Boston: Routledge & Kegan Paul, 7–27.

Palloix, C. (1977) 'The self-expansion of capital on a world scale', *The Review of Radical Political Economics (URPE)*, 9, 2, 1–28.

Palma, G. (1978) 'Dependency: a formal theory of under-development or a methodology for the analysis of concrete situations of under-development?', *World Development*, 6, 7 & 8, 881–922.

Parboni, R. (1980) *The Dollar and its Rivals*, London, Verso (New Left Books).

Payer, C. (1982) *The World Bank: A Critical Analysis*, New York and London, Monthly Review Press.

Pearse, Andrew (1980) *Seeds of Plenty, Seeds of Want: Social and Economic Implications of the Green Revolution*, London, Oxford University Press.

Petras, J. (1975) 'New perspective on imperialism and social classes in the periphery', *Journal of Contemporary Asia*, 5, 3, 291–308.

Pinto, A. and Kñakal, J. (1972) 'El sistema centro-periferia veinte años después', *Revista de la Integracion*, 10, 5–82.

Portes, A. and Walton, J. (1981) *Labor, Class and the International System*, New York, Academic Press.

Poulantzas, N. (1974) 'Internationalization of capitalist relations and the nation-state', *Economy and Society*, 3, 2, 145–79.

Poulantzas, N. (1975) *Classes in Contemporary Capitalism*, London, New Left Books.

Radice, H. (ed.) (1975) *International Firms and Modern Imperialism*, Harmondsworth, Penguin.

Raffestin, C. and Bresso, M. (1979) *Travail, espace et pouvoir*. Lausanne, Editions l'Age d'Homme.

Roberts, B. (1978) *Cities of Peasants: The Political Economy of Urbanization in the Third World*, London, Edward Arnold.

Rodriguez, O. (1979) 'La teoría del subdesarrollo de la Cepal: síntesis y critica', *Comercio Exterior*, 29, 11, 1177–93.

Roncagliolo, R. and Janus, N. (1981) 'Advertising and the democratization of communications', *Development Dialogue*, 2, 31–40.

Roxborough, I. (1979) *Theories of Underdevelopment*, London, Macmillan.

Safa, H. (1980) 'Export processing and female employment: the search for cheap labour', paper presented for Burg Wartenstein Symposium 85, Werner-Gren Foundation for Anthropological Research.

Salih, Kamal (1980) *Report on Intermediate Cities Conference*, Honolulu, East-West Center.

—— (1981) 'Territorial social formation: notes on some concepts underlying a strategy of regional closure', paper presented at workshop on geographic transfer of value, Canberra, Australian National University, December.

Sampson, Anthony (1981) *The Money Lenders: Bankers in a Dangerous World*, London, Hodder & Stoughton, Coronet Books.

Sandbrook, R. (1982) *The Politics of Basic Needs*, Toronto, University of Toronto Press.

Santos, M. (1979) *The Shared Space: The Two Circuits of the Urban Economy in Underdeveloped Countries*, London and New York, Methuen.

Scott, J. W. and Tilly, L. A. (1981) 'Women's work and the family in nineteenth century Europe', in M. Anderson (ed.) *Sociology of the Family*, Harmondsworth, Penguin, 126–63.

Shanin, T. (ed.) (1971) *Peasant and Peasant Societies*, Harmondsworth, Penguin.

Sharpston, M. (1973) 'International sub-contracting', *Oxford Economic Papers*, New Series, 27, 1, 94–135.

Singh, A. (1979) 'The basic needs approach to development versus the new international economic order: the significance of Third World industrialization', *World Development*, 7, 585–606.

Sklar, H. (1980) *Trilateralism: The Trilateral Commission and Elite Planning for World Management*, Montreal, Black Rose Books.

Slater, D. (1979) 'Imperialism and the limitations on capitalist transformation at the periphery', in J. Carriére (ed.) *Industrialisation and the State in Latin America*, Amsterdam, Cedla Publications, 91–126.

Soja, E. *et al.* (1983) 'Urban restructuring: an analysis of social and spatial change in Los Angeles', *Economic Geography*, 59, 2, 195–230.

Somavía, J. (1981) 'The democratization of communications: from minority social monopoly to majority social representation', *Development Dialogue*, 2, 13–30.

Squire, Lyn (1979) *Labor Force, Employment and Labor Markets in the Course of Economic Development*, World Bank Staff Working Paper, 336, Washington, DC, International Bank of Reconstruction.

Tabb, W. K. and Sawers, L. (1978) *Marxism and the Metropolis*, London, Oxford University Press.

Taylor, J. (1979) *From Modernization to Modes of Production*, London, Macmillan.
Thompson, E. P. (1978) *The Poverty of Theory and other Essays*, London, Merlin Press.
—— (1981) 'The politics of theory', in R. Samuel (ed.) *People's History and Socialist Theory*, London, Routledge & Kegan Paul, 396–408.
Time Magazine (1983) 'The debt threat', 10 January, 32–41.
Topalov, C. (1979) *La Urbanización Capitalista: Algunos Elementos para su Análisis*, Mexico City, Editorial Edicol.
Torchinsky, R. (1980) 'Some theoretical and spatial aspects of manufactured exports from less developed countries', unpublished paper (limited circulation).
Tucker, R. C. (1978) *The Mark-Engels Reader*, New York, W. W. Norton.
Ul Haq, M. (1976) *The Poverty Curtain: Choices for the Third World*, New York, Columbia University Press.
UNCTAD (1978) 'Review of recent trends and developments in trade and semi-manufacturers', Geneva, UNCTAD, TD/B/C 2, 190.
Urencio, C. C. F. (1979) 'La estratégia trilateral y los países en desarrollo', *Comercio Exterior*, 29, 11, 1206–11.
Wallerstein, I. (1974) *The Modern World-System*, New York, Academic Press.
—— (1976) 'Semi-peripheral countries and the contemporary world crisis', *Theory and Society*, 3, 4, 461–83.
Warren, B. (1973) 'Imperialism and capitalist industrialization', *New Left Review*, September-October.
—— (1975) 'How international is capital?', in H. Radice (ed.) *International Firms and Modern Imperialism*, Harmondsworth, Penguin, 135–40.
Weeks, J. (1971) 'Does employment matter?' *Manpower and Unemployment Research in Africa*, 4, 1, 67–70.
Wolfe, H. (ed.) (1980) *The Articulation of Modes of Production*, London, Routledge & Kegan Paul.
Wolfe, M. (1976) *El Desarrollo Esquivo: Exploraciones en la Política Social y la Realidad Socio-Política*, Mexico City, Cepal/Fondo de la Cultura Económica.
Zaretsky, E. (1976) *Capitalism, the Family and Personal Life*, London, Pluto Press.
Zukav, GB. (1979) *The Dancing Wu Li Masters: An Overview of the New Physics*, Toronto and New York, Bantam Books.

Latin America

Abalo, C. (1981) 'Argentina, 1976–1981: objetivos y resultados de la política económica', *Comercio Exterior*, 3, 6, 635–48 (Part 1); 3, 9, 1017–30 (Part 2).
Agee, P. (1975) *Inside the Company: CIA Diary*, Harmondsworth, Penguin.
Alborñoz, P. O. (1976) *Las Luchas Indígenas en el Ecuador*, Guayaquil, Editorial Claridad.
Armstrong, R. (1980) 'El Salvador', *Nacla Report on the Americas*, XIV, 2, 2–35.
Armstrong, W. R. (1973) 'Crítica de la teoría de polos de desarrollo', *Revista Latino-Americana de Estudios Urbanos y Regionales (Eure)*, Santiago, 3, 7, 113–23.
—— (1974) 'Evaluación del sistema de planificación regional en Ecuador' (Project report on regional planning), Quito, Naciones Unidas/Junapla, Proyecto ECU/74/005.
—— (1975) 'Población, medio ambiente y desarrollo: la experiencia Latino-Americana, Comisión Económica para America Latina (CEPAL), *Población y Desarrollo en América Latina*, México, Fondo de la Cultura Ecónomica.

Armstrong, W. R. (1978) 'Development as a smoke-screen: the worth of a United Nations project', *Working Paper Series*, 21, CDAS (McGill University).

Baird, P. and McCaughan, E. (1979) *Beyond the Border: Mexico and the U.S. Today*, New York, Nacla.

Bambirra, V. (1973) *Capitalismo Dependiente Latino-Americano*. Santiago, CESO/ Editorial Prensa Latino-Americana.

Barraclough, S. and Collarte, J. C. (1973) *Agrarian Structure in Latin America*, Lexington, D. C. Heath.

Barry, T., Wood, B. and Preusch, D. (1982) *Dollars and Dictators: A Guide to Central America*, Albuquerque, The Resource Center.

Bastias, A. C. (1980) 'Chile 1973–1980: la nueva estratégia de desarrollo y su aplicación', *Comercio Exterior*, 30, 9.

Black, G. (1983) 'Garrison Guatemala', *Nacla Report*, XVII, 1, 2–35.

Blackstein, C. S. and Zuvekas, C. (1973) 'Agrarian reform in Ecuador: evaluation of past efforts and the development of a new approach', *Economic Development and Cultural Change*, 22, 1, 73–94.

Bocco, A. (1982) 'Ecuador: economic policy and development promotion during the petroleum boom: 1972–78', *Northsouth*, VII, 14, 43–66.

Borja, R. (1980) 'Ecuador: underdeveloping democracy', *Nacla Report on the Americas*, XIV, 6, 36–9.

Bravo, G. (n.d.) *Los Albañiles en Cuenca*, Cuenca, Instituto de Investigaciones Sociales de la Universidad de Cuenca, Separata 3.

Brodersohn, M. (1971) 'Regional development and industrial location policy in Argentina', New York, UNIDO, United Nations, 474–552.

Bromley, R. J. (1977) *Development Planning in Ecuador*, London, Latin American Publications Fund.

Burbach, R. and Flynn, P. (1980) *Agribusiness in the Americas*, New York, Monthly Review Press.

Calcagno, A. E. and Jakobowicz, J. M. (1981) 'Some aspects of the international distribution of industrial activity', *Cepal Review*, 13, 7–34.

Cardenal, E. (1972) 'Economía de Tahuantinsuyo', *Homenaje a los Indios Americanos*, Buenos Aires, Cuadernos Americanos.

Carrasco A. (n.d.) *Evolución y Estructura de las Organizaciones de Trabajadores en el Azuay*. Cuenca, Universidad de Cuenca, Separata 2.

CEPAL Review (1982) Edition on Food, Agriculture and the Peasantry in Latin America, *Cepal Review*, 16, 1–176.

CEPAL/United Nations (1969) *Economic Development and Income Distribution in Argentina*, New York, United Nations.

—— —— (1974) *Statistical Yearbook for Latin America 1973*, vols I and II, New York, United Nations/CEPAL.

—— —— (1977) *Statistical Yearbook for Latin America 1976*, New York, United Nations/CEPAL.

—— —— (1978) *Series Históricas del Crecimiento de América Latina*, Santiago, CEPAL.

—— —— (1980) *Statistical Yearbook for Latin America 1979*, New York, United Nations/CEPAL.

Collier, D. (1978) 'Industrial modernization and political change', *World Politics*, XXX, 4.

Collier, D. (ed.) (1979) *The New Authoritarianism in Latin America*, Princeton, NJ, Princeton University Press.

Comercio Exterior (1979a) 'Recuento Latino Americano: conferencia Latino-Americana de asentamientos humanos', *Comercio Exterior*, 29, 11, 1230.

—— (1979b) 'Asentamientos humanos: vialidad y transporte en la gran urbe'. *Comercio Exterior*, 29, 10, 1083–9.

—— (1981) 'Brasil: económía y política: dos tropiezos del plan del Gobierno', *Comercio Exterior*, 31, 3, 269–74.

Cordero, L. C. (1976) *El Mercurio* (Cuenca), 31 May.

Cornelius, W. A. and Kemper, R. V. (eds) (1977) 'Metropolitan Latin America: the challenge and the response', *Latin American Urban Research*, 6, Beverly Hills, Sage.

—— and Trueblood, F. M. (eds) (1975) *Latin American Urban Research, vol. 5: Urbanization and Inequality: The Political Economy of Urban and Rural Development in Latin America*, Beverly Hills, Sage.

Corradi, J. E. (1974) 'Argentina', in R. H. Chilcote and J. C. Edelstein (eds) *Latin America: The Struggle with Dependency and Beyond*, New York, Halsted Press/John Wiley & Sons, 305–408.

Cortes Conde, R. and Gallo, E. (1967) *La Formación de Argentina Moderna*, Buenos Aires, Paidos.

Cotler, J. (1979) 'State and regime: comparative notes on the southern cone and the "enclave" societies', in D. Collier (ed.) *The New Authoritarianism in Latin America*, Princeton, NJ, Princeton University Press, 255–82.

CREA (1976) *CREA: Desarrollo Regional 1975: Informe*, Cuenca, CREA.

—— (1977) *Plan del Desarrollo Regional 1978–82 para las Provincias del Azuay, Cañar y Morona Santiago* (vol. 1, Caracterización del desarrollo regional; vol. 2, Diagnóstico por areas basicas de planificación). Cuenca, CREA.

Diaz Alejandro, C. F. (1970) *Essays on the Economic History of the Argentina Republic*, New Haven, Conn., Yale University Press.

Dodero, C. H. (1979) 'El nuevo orden económica internacional: la promesa y la realidad', *Comercio Exterior*, 29, 11, 1212–20.

Dorfman, A. (1970) *Historia de la Industria Argentina*, Buenos Aires, Solar-Hachette.

Economic Commission for Latin America (ECLA) (1961) *Economic Development, Planning and International Cooperation*, Santiago, Economic Commission for Latin America.

—— (1967) *Estudio sobre la Distribución del Ingreso en América Latina*, Santiago, United Nations.

El Mercurio, Cuenca (1976) article on urban delinquency, 9 June.

Espinosa, S. (1976) 'Nuestra Iglesia: una Iglesia entrampada', *El Mercurio*, Cuenca, 6 June.

Evans, P. (1977) 'Multinational, state-owned corporations and the transformation of imperialism: a Brazilian case study', *Economic Development and Cultural Change*, 26, 1.

—— (1979) *Dependent Development: The Alliance of Multinational, State and Local Capital In Brazil*, Princeton, NJ, Princeton University Press.

Feder, E. (1977) 'Agribusiness and the elimination of Latin America's rural proletariat', *World Development*, 5, 5–7, 559–71.

Ferrer, A. (1967) *The Argentine Economy*, Berkeley, University of California Press.

—— (1981) 'El monetarismo en Argentina y Chile', *Comercio Exterior*, 31, 1, 3–13 (part 1); 31, 2, 176–92 (part 2).

Flynn, P., Santana, A. and Shapiro, H. (1980) 'Latin American women: one myth – many realities', *NACLA Report on the Americas*, XIV, 5, 2–35.

Fortin, C. (1979) 'The state and capital accumulation in Chile', in J. Carriére (ed.) *Industrialisation and the State in Latin America*, Amsterdam, Centre for Latin American Research and Documentation, 15–48.

Fox, R. W. (1975) *Urban Population Growth Trends in Latin America*, Washington, DC, Inter-American Development.

Foxley, A. (ed.) (1976) *Income Distribution in Latin America*, London, Cambridge University Press.

Frank, A. G. (1967) *Capitalism and Underdevelopment in Latin America: Historical Studies of Chile and Brazil*, New York, Monthly Review Press.

Furtado, C. (1970) *Economic Development of Latin America: A Survey from Colonial Times to the Cuban Revolution*, Cambridge, Cambridge University Press.

Galarza, J. Z. (1975) *El Yugo Feudal: Visión del Campo Ecuatoriano*, Quito, Ediciones Solitierra.

Garavaglia, J. C. (1973) 'Introducción', in C. Assadourian *et al. Modos de Producción en América Latina*, Cordoba (Argentina), Cuadernos de Pasado y Presente, 40.

Geisse, G. and Coraggio, J. L. (eds) (1972) 'Metropolitan areas and national development', in G. Geisse and J. E. Hardoy (eds) *Latin American Urban Research, 2*, Beverly Hills, Sage, 45–59.

Gilbert, A. G. (1980) 'Planning for urban primacy and large cities in Latin America: a critique of the literature', *Comparative Urban Research*, VIII, 1, 105–16.

Gilbert, A. (1982) 'Mining and manufacturing', in B. W. Blouet and O. M. Blouet (eds), *Latin America: An Introductory Survey*, New York, John Wiley, 250–77.

Girault, C. (1981) 'La Gulf and Western en la República Dominicana: del enclave azucarero al control parcial de la economía', *Comercio Exterior*, 31, 3, 278–83.

Government of Ecuador (1972) *Filosofía y Plan de Acción del Gobierno Revolucionario y Nacionalista del Ecuador*, Quito (mimeo).

Graham, D. (1970) 'Divergent and convergent regional economic growth and internal migration in Brazil', *Economic Development and Cultural Change*, 18, 362–82.

Griffin, K. B. (1969) *Underdevelopment in Spanish America: An Interpretation*, London, Allen & Unwin.

—— (1974) 'Systems of labour control and rural poverty in Ecuador', Oxford, Prealc, ILO (mimeo).

Gutman, P. (1982) 'Problemas y perspectivas ambientales de la urbanización en América Latina', *Comercio Exterior*, 32, 12, 1304–14.

Hardoy, J. E. (ed.) (1975) *Urbanization in Latin America: Approaches and Issues*, New York, Anchor Books.

Henfrey, C. (1981) 'Dependency, modes of production, and the class analysis of Latin America', *Latin American Perspectives*, VIII, 3 & 4, 17–54.

Horowitz, D. (1970) 'The alliance for progress', in R. I. Rhodes (ed.) *Imperialism and Underdevelopment: A Reader*, New York, Monthly Review Press, 45–61.

Huizer, G. (1973) *Peasant Rebellion in Latin America*, Harmondsworth, Penguin.

Icaza, J. (1953) *Huasipungo*, Buenos Aires, Editorial Losada.

Jaramillo, P. A. (1954) *El Indio Ecuatoriano*, Quito, Casa de la Cultura Ecuatoriana (first published 1922).

Jelin, E. (1979) 'Labour conflicts under the second Peronist regime, Argentina 1973–76', *Development and Change*, 10, 233–57.

Jenkins, R. (1979) 'Transnational corporations and their impact on the Mexican economy', in J. Carriére (ed.) *Industrialisation and the State in Latin America*, Amsterdam, CEDLA Publications, 157–88.

Jorge, E. (1971) *Industria y Concentración Económica: desde Principios del Siglo hasta el Peronismo*, Buenos Aires, Siglo XXI.

Junta Nacional de Planificación (1972) *Ecuador: Plan Integral de Transformación y Desarrollo, 1973–1977: Resumen General*, Quito, Editorial 'Santo Domingo'.

—— (1975) *III Censo de Población: II Censo de Vivienda, 1974*, Quito, Oficina de los Censos Nacionales.

Juruna, J. (1981) 'Brésil: la dette extérieure; facteur de déstabilisation', *Le Monde Diplomatique*, 333, 14.

Kaplan, D. (1965) 'The Mexican market place then and now', *Proceedings of the 1965 Annual Spring Meeting of the American Ethnological Society*, 80–94.

Keeler, R. H., Mera, R. and Cruz, R. P. (1973) 'Evaluation of the directed agricultural credit program in Ecuador', *USAID Spring Review of Small Farmer Credit*, IV, 4.

Laclau, E. (1971) 'Feudalism and capitalism in Latin America', *New Left Review*, May-June, 19–38.

Lahera, E. (1981) 'The transnational corporations in the Chilean economy', *Cepal Review*, 14, 107–26.

Lain, E. P. (1966) 'Buenos Aires', *Cuaderno Cultural* (Spain), 4, 6, 19–25.

Landsberg, M. (1979) 'Export-led industrialization in the Third World', *The Review of Radical Political Economics*, II, 4, 50–63.

Larrea-Maldonado, C. (1982) 'Transnational companies and banana exports from Ecuador, 1948–1972: an interpretation', *Northsouth*, VII, 14, 3–42.

Latin American Economic Report (1979a) 'El Salvador government hesitates over strike wave', *LAER*, VII, 15.

—— (1979b) 'New regime faces labour trouble in El Salvador', *LAER*, VII, 42.

Ledogar, R. J. (1976) *Hungry for Profits: US Food and Drug Multinationals in Latin America*, New York, IDOC.

Malloy, J. M. (ed.) (1977) *Authoritarianism and Corporatism in Latin America*, Pittsburgh, University of Pittsburgh Press.

Mangin, W. P. (1967) 'Latin American squatter settlements: a problem and a solution', *Latin American Research Review*, 2.

Marinho, L. C. (1981) 'The Transnational corporations and Latin America's present form of economic growth', *Cepal Review*, 14, 9–34.

Martel, J. (1978) 'Domination by debt: finance capital in Argentina', *NACLA Report on the Americas*, XII, 4, 20–38.

Monsalve, L. P. (1943) *El Indio: Cuestiones de su Vida y su Pasión*, Cuenca, Editorial Austral.

Morse, R. M. (1975a) 'The city-idea in Argentina: a study in evanescence', *Revista da Historia* (Brazil), 52, 104, 783–802.

—— (1975b) 'The development of urban systems in the Americas in the nineteenth century', *Journal of Interamerican Studies and World Affairs*, 17, 1, 4–26.

—— (1980) 'Urban development in Latin America: introduction', *Comparative Urban Research*, VIII, 1, 5–13.

Movarec, M. (1982) 'Exports of Latin American manufactures to the centres: their magnitude and significance', *Cepal Review*, 17, 47–77.

Munck, R. (1979) 'State and capital in dependent social formations: the Brazilian case', *Capital and Class*, 8, 34–53.

Navarro, G. (1976) *La Concentración de Capitales en el Ecuador*, Quito, Ediciones Solitierra.

Nun, J. (1969) 'Sobrepoblación relativa, ejército industrial de reserva y marginal', *Revista Latinoamericana de Sociologia*, 4, 2, 178–237.

Ominami, C. (1981) '"Libéralisation" au Chili: retour en force des capitaux étrangers', *Le Monde Diplomatique*, January.

Petras, J. and Laporte, R. (1973) *Cultivating Revolution: The US and Agrarian Reform in Latin America*, New York, Random House (Vintage Books).

Platt, D. C. M. (1977) *Business Imperialism, 1840–1930: An Enquiry Based on British Experience in Latin America*, Oxford, Clarendon Press.

Portes, A. and Browning, H. (eds) (1976) *Current Perspectives in Latin American Urban Research*, Austin and London, University of Texas Press.

Quijano, A. (1975) 'The urbanization of Latin American society', in J. Hardoy (ed.) *Urbanization in Latin America: Approaches and Issues*, New York, Anchor Books, 109–53.

Rofman, A. B. (1974) *Dependencia, Estructura de Poder y Formación Regional en América Latina*, Buenos Aires, Siglo XXI.

Ruiz Nápoles, P. (1981) 'El comercio entre paises desarrollados y subdesarrollados: algunos aspectos teóricos y de la política económica', *Comercio Exterior*, 31, 10, 1173–8.

Sargent, C. S. (1974) *The Spatial Evolution of Greater Buenos Aires, 1870–1930*, Tempe, Arizona State University.

—— (1982) 'The Latin American City', in B. W. Blouet and O. M. Blouet (eds) *Latin America: An Introductory Survey*, New York, John Wiley, 201–49.

Scobie, J. R. (1972) 'Buenos Aires as a commercial–bureaucratic city, 1880–1910: characteristics of a city's orientation', *American History Review*, 77, 24, 1035–73.

—— (1975) 'Argentina: patterns of urbanization in Argentina, 1869–1914', *Latin American Research Review*, 10, 2, 132–4.

Slater, D. (1978) 'Towards a political economy of urbanization in peripheral capitalist societies: problems of theory and method with illustrations from Latin America', *International Journal of Urban and Regional Research*, 2, 1, 26–52.

Soza, H. (1981) 'The industrialization debate in Latin America', *Cepal Review*, 13, 35–64.

Stein, S. and Stein, B. (1970) *The Colonial Heritage of Latin America: Essays on Economic Dependence in Perspective*, New York, Oxford University Press.

Tavares, M. da C. (1979) *De la Sustitución de Importaciones al Capitalismo Financiero*, Mexico, Fondo de la Cultura Económica.

—— and Teixeira, A. (1981) 'Transnational enterprises and the internationalization of capital in Brazilian industry', *Cepal Review*, 14, 85–106.

Tax, S. (1953) *Penny Capitalism: A Guatemalan Indian Economy*, Institute of Social Anthropology, Smithsonian Institution Publication 16, Washington, DC, Smithsonian Institution.

UNESCO, CEPAL, PNUD (1981) *La Educación y los Problemas del Empleo*, Proyecto Desarrollo y Educación en América Latina y el Caribe, Buenos Aires, CEPAL.

UNOTC (United Nations Office of Technical Cooperation) (1974) 'Evaluación del

sistema de planificación regional', Quito, ECU/74/005/Junta Nacional De Planificación (mimeo).
Vapñarsky, C. A. (1975) 'The Argentine system of cities: primary and rank-size rule', in: J. E. Hardoy (ed.) *Urbanization in Latin America: Approaches and Issues*, New York, Anchor Books, 369–90.
Velasco, F. (1973a) 'El modelo agro-exportador Ecuatoriano', *Economía* (Universidad Central, Quito), January.
—— (1973b) 'Notas sobre la dependencia Ecuatoriana', *Economía* (Universidad Central, Quito), June.
Vicuña, C. I. (1975) *La Clase Trabajadora del Ecuador: Características y Condiciones de Vida*, Guayaquil, Universidad de Guayaquil.
Vilas, C. M. (1982) 'On the rise of bourgeois democracy in Argentina', *Latin American Perspectives*, 35, IX, 4, 5–30.
'Voz del Campesino' (1976) 'El CREA para los marginados', *El Mercurio* (Cuenca), 5 June.
Walton, J. (1976a) 'Political economy of world urban systems: directions for comparative research', in J. Walton and L. H. Massoti (eds) *The City in Comparative Perspective: Cross-National Research and New Directions in Theory*, Los Angeles, Sage.
—— (1976b) 'Urban hierarchies and patterns of dependence in Latin America: theoretical bases for a new research agenda', in A. Portes and H. L. Browning (eds) *Current Perspectives in Latin American Urban Research*, Austin, University of Texas Press.
—— (1977) *Elites and Economic Development: Comparative Studies on the Political Economy of Latin American Cities*, Austin and London, University of Texas Press.
Wells, J. (1977) 'The diffusion of durables in Brazil and its implications for recent controversies concerning Brazil and development', *Cambridge Journal of Economics*, September, 259–80.
Wilber, J. W. and Haber, S. (eds) (1978) *Statistical Abstract of Latin America*, vol. 21, Los Angeles, University of California at Los Angeles Latin American Center Publications (from *Series Históricas de Crecimiento de América Latina*, Santiago, Comisión Económica para América Latina, 74ff.).
Wingo, L. (1967) 'Recent patterns of urbanization among Latin American countries', *Urban Affairs Quarterly*, 2, 81–110.
World Bank (IBRD) (1981) *World Development Report 1981*, New York, World Bank/Oxford University Press.
Wynia, G. W. (1978) *Argentina in the Post-War Era: Politics and Economic Policy-Making in a Divided Society*, Albuquerque, University of New Mexico Press.

Asia

AMPO (1977) *Free Trade Zones and Industrialization in Asia*, Tokyo, Pacific Asia Resources Center.
Ariffin, J. A. (1980) 'Female labour migration to urban-based factories in Malaysia and Malay women's participation. The labour force: an analysis of the changing trends and patterns', unpublished paper presented at the Eleventh Summer Session Workshop on 'Analysis of Female Migration', East-West Center, Honolulu.
—— (1981) 'Women at work', *Third Eye*, Kota Kinabalu, 12–22.

Ariffin, J. A. (1982) 'Industrialization, female labour migration and the changing pattern of Malay women's labour force participation', *Southeast Asian Studies*, 19, 4, 412–25.
Asian Development Bank (1982) *Key Indicators of Developing Member Countries of ADB*, Manila, Asian Development Bank.
Chan, Tuck Hoong and Kok, Kim Lian (1984) 'Urban development in Malaysia: trends, patterns and policy issues', paper presented to the International Conference on Urban Growth and Economic Development in the Pacific Region, Institute of Economics, Academia Sinica, Taipei, Taiwan.
Chen, E. K. (1980) 'The business environment in Hong Kong', in P. G. Lethbridge (ed.) *The Business Environment in Hong Kong*, Hong Kong, Oxford University Press, 1–50.
Choo, C. S. (1977) 'The impact of foreign manufacturing firms on the local region: a case study of Prai Free Trade Zone, Penang', *Geographica*, 12, 21–30.
Collier, William *et al.* (1982) 'The acceleration of rural development on Java: from village studies to national perspective', *Agro-Economic Survey*, Rural Dynamics Study, Occasional Paper, Bogor.
Cushman, R. E. (1976) 'Sources and uses of capital in the Asian region', *Euromoney*, May supplement, 13.
Davis, W. G. (1968) 'Economic limitations and social relationships in a Philippino marketplace. Capital accumulation in a peasant economy', in R. Van Niel (ed.) *Economic Factors in Southeast Asian Social Change*, Hawaii, University of Hawaii.
De Glopper, D. R. (1972) 'Doing business in Lukang', in W. E. Willmont (ed.) *Economic Organization in Chinese Society*, Stanford, Calif., Stanford University Press, 297–326.
De Koninck, R. (1981) 'Work, space and power in the rice fields of Kedah: reflections on the dispossession of territory', unpublished paper presented at the Conference of the Canadian Council of Southeast Asian Studies, University of Calgary, November.
Dewey, A. (1962) *Peasant Marketing in Java*, Glencoe, Ill., Free Press.
Dotson, A. T. (1972) 'Urbanization and development in Southeast Asia. The urban development panel seminars: 1971–72', Washington, DC, Seadag Reports.
Drakakis-Smith, D. W. (1970) 'Hong Kong housing preferences and housing policy', Centre of Asian Studies, University of Hong Kong (mimeo).
Economic and Social Commission for Asia and the Pacific (1978) *Economic Bulletin for Asia and the Pacific*, Economic and Social Commission for Asia and the Pacific, XXVIII, 1 & 2.
—— (1980) *Migration, Urbanization and Development in the Republic of Korea*, Bangkok, ESCAP, Population Division.
Forbes, D. (1982) 'Energy imperialism and a new international division of labour', *Tijdschrift voor Economische en Social Geografie*, LXXIII, 2, 94–108.
Friedmann, J. P. and Douglass, M. (1976) 'Agropolitan development: towards a new strategy for regional planning in Asia', in United Nations Centre for Regional Development, *Growth Pole Strategy and Regional Planning Development in Asia*, Nagoya, UNCRD, 333–87.
Goldstein, Sidney (1978) 'Circulation in the context of total mobility in Southeast Asia', paper prepared for the International Seminar on the Cross-Cultural Study of Circulation, East-West Center, Honolulu, Hawaii.
Golley, Linda (1983) 'Japanese capital in Malaysia: investment paradise', *Southeast Asia Chronicle*, 88, 14–18.

Grossman, R. (1978) 'Women's place in the integrated circuit factories', *Pacific Research*, IX, 5 & 6, 2–17.

—— (1979) 'Women's place in the integrated circuit', *Southeast Asia Chronicle*, 66, 2–17.

Hauser, P. M. (ed.) (1957) *Urbanization in Asia and the Far East*, Tensions and Technology Series, Calcutta, UNESCO.

Hauser, P. M. (1972) 'Population change and developments in manpower, labour force, employment and income, paper prepared for UNECAFE Seminar on Population Aspects of Social Development, Bangkok.

Hirschman, C. (1976) 'Recent urbanization trends in peninsular Malaysia', *Demography*, 13, 445–62.

Ho, K. P. (1979) 'Birth of the second generation', *Far Eastern Economic Review*, 14, 20, 12–14.

—— (1980) 'Bargaining on the Free Trade Zones', *New Internationalist*, March.

Ho, S. F. (1972) 'Hawkers in Mongkok District: a study of retailing geography', unpublished M.Phil. thesis, Department of Geography, University of Hong Kong.

Hong Kong Government Mongkok City District Office (1968) 'Report on the hawker situation in Mongkok', unpublished report.

Hong Kong Government Resettlement Department (1969) 'Hawker survey', unpublished report.

Hong Kong Government (1972) *Hong Kong Population and Housing Census, 1971*, Hong Kong, Hong Kong Government Printer.

Hong Kong Government, Department of Statistics (1983) *Hong Kong Monthly Digest of Statistics*, January.

Hong Kong Urban Council and Urban Services Department, *Annual Departmental Report, 1946–47*, Hong Kong, Hong Kong Government Printer.

Hong Kong Urban Council Proceedings (1958 . . .), Hong Kong, Hong Kong Government Printer.

Hopkins, K. (1969) 'Public housing policy in Hong Kong. An inaugural lecture from the Chair of Sociology', *University of Hong Kong, supplement to the Gazette*, XVI, 1–12.

Hugo, Graeme (1975) 'Population mobility in West Java', unpublished Ph.D. thesis in demography, Australian National University, Canberra.

Johnstone, M. A. (1981) 'The evolution of squatter settlements in Peninsular Malaysian cities', *Journal of Southeast Asian Studies*, 12, 2, 364–80.

Jones, G. (1965) 'The employment characteristics of small towns in Malaya', *Malayan Economic Review*, 10, 1, 44–72.

Kassalow, E. M. (1978) 'Aspects of labour relations in multinational companies: an overview of three Asian countries', *International Labour Review*, 117, 3, 273–89.

Kojima, K. (1978) *Japanese Direct Foreign Investment, A Model of Multinational Business Operations*, Tokyo, Charles E. Tuttle.

Koo, H. (1980) 'Urbanization pattern and emerging inequality structure in Korean development, 1960–1979', unpublished paper.

Kulkarni, V. G. (1983) 'An unreal estate boom', *Far Eastern Economic Review*, 8 September, 71–8.

Lagbao, F. Y. and Pa, R. E. (1981) 'Transnationals and consumerism: impact on Filipino tastes and values', *State and Society*, 2, 1, 91–9.

Lee, S. Y. and Jao, V. C. (1982) *Financial Structures and Monetary Policies in Southeast Asia*, London, Macmillan.

Lim, L. Y. C. (1980) 'Women in the redeployment of manufacturing industry to developing countries', *UNIDO Working Papers on Structural Change*, 18.
Lim, Lin Lean (1982) 'Towards meeting the needs of urban female factory workers in peninsular Malaysia', paper presented to the Conference on Women in the Urban and Industrial Workforce, Southeast and East Asia, Manila, University of the Philippines.
Lim, S. G. L. (1982) 'Mr Tang's Girls', *Asiaweek*, October 1, 42–51.
Lo, C. P. (1972) 'A typology of Hong Kong census districts: a study in urban structure', *The Journal of Tropical Geography*, 34, June, 34–43.
Lo, F. C. and Salih, K. (eds) (1978) *Growth Pole Strategy and Regional Development Policy: Asian Experience and Alternative Approaches*, Oxford, Pergamon Press.
—— —— (1979) 'Growth poles, agropolitan development and polarization reversal: the debate and research for alternatives', *United National Centre for Regional Development Working Papers*, 79–108.
Lu, A. L. C. (1972) *Hawkers and Their Relocation in Hong Kong*, Hong Kong, Social Research Centre, The Chinese University.
McGee, T. G. (1967) *The Southeast Asian City*, London, G. Bell & Son.
—— (1969) *Malays in Kuala Lumpur City*, unpublished Ph.D. thesis, Victoria University of Wellington, New Zealand.
—— (1970) *Hawkers in Selected Southeast Asian Cities*, Hong Kong, Centre for Asian Studies, University of Hong Kong.
—— (1971a) 'Têtes-de-Ponts et enclaves: le problème urbain et le processus d'urbanization dans l'Asie du Sud-est depuis 1945', *Revue Tiers Monde*, XII, 45, 115–44.
—— (1971b) 'Catalysts or cancer? The role of cities in Asian society', in L. Jacobson and V. Prakash (eds) *Urbanization and National Development*, Beverly Hills, Calif., Sage, 157–81.
—— (1974) *Hawkers in Hong Kong. A Study of Policy and Planning in the Third World City*, Hong Kong, Centre for Asian Studies, University of Hong Kong.
—— (1978a) 'An invitation to the ball: dress formal or informal?', in P. J. Rimmer, D. W. Drakakis-Smith, and T. G. McGee (eds) *Food, Shelter and Transport in Southeast Asia and the Pacific*, Publication HG/12, Canberra, Australian National University.
—— (1978b) 'Rural-urban mobility in South and Southeast Asia: different formulations, different answers', in W. H. McNeil and R. S. Adams (eds) *Human Migration: Patterns and Policies*, Bloomington, Indiana University Press.
—— (1979a) 'Conservation and dissolution in the Third World city: the shanty town as an element of conservation', *Development and Change*, 10, 1, 1–22.
—— (1979b) *Labour Mobility in Fragmented Labour Markets: Rural-Urban Linkages and Regional Development in Asia*, URCRD Working Paper 79–105, Nagoya, URCRD.
—— (1981a) 'Labour markets, urban systems and the urbanization process in Southeast Asia', unpublished paper presented to the Seminar on Secondary Cities in Asia, East-West Center, Honolulu.
—— (1981b) 'The urbanization process and industrialization in the market economies of Asia', *Regional Development Dialogue*, 1, 2, 131–43.
—— (1982) 'Proletarianization, industrialization and urbanization in Asia: a case study of Malaysia', Flinders Asian Studies Lecture, 13, Adelaide, Flinders University.
—— (1985a) 'Circuits and networks of capital: the internalization of the world economy and national urbanization' in *Conference on Urban Growth and Economic Development in the Pacific Region*, Taipei, The Institute of Economics Academia Sinica, 81–100.
—— (1985b) 'From "urban involution" to "proletarian transformation": Asian perspectives', in T. G. McGee (ed.) *Proletarianization in Asia and the Pacific*, Centre for

Southeast Asian Studies Monograph, Monash University, Melbourne.
—— and Yeung Yue-man (1977) *Hawkers in Southeast Asian Cities*, Ottawa, IDRC.
McGee, T. G., Ward, R. and Drakakis-Smith, D. (1980) *Food Distribution in the New Hebrides*, Monograph 25, Development Studies Centre, ANU, Canberra.
Marshall, G. A. (1964) 'Women, trade and the Yoruba family', unpublished Ph.D. thesis, New York, Columbia University.
Musaffar, C. (1979) *Protector?*, Penang, Alisan.
O'Brien, L. N. (1981) 'Asian women and the new international division of labour', paper presented to the workshop 'Women in Asia', Asian Studies Association of Australia, University of New South Wales, June.
—— (1982) 'The effect of industrialization on women: western and current Southeast Asian experiences', paper presented at the Conference on Women in the Urban and Industrial Workforce, Southeast and East Asia, Manila, University of the Philippines.
Oriental Economist (Shukan Toyokeizai) (1982) Tokyo, 9 December.
Penang Development Corporation (1980) *Annual Report*, Penang.
Pryor, E. G. (1971) 'An assessment of the needs and scope for urban renewal in Hong Kong, unpublished Ph.D. thesis, Department of Geography, University of Hong Kong.
Pryor, R. J. (ed.) (1979) *Migration and Development in Southeast Asia: A Demographic Perspective*, Kuala Lumpur, Oxford University Press.
Puthucheary, J. J. (1960) *Ownership and Control in the Malaysian Economy*, Singapore, Eastern Universities Press.
Robinson, K. (1985) 'The wedding of Hijra: changes in marriage and social relations in an Indonesian mining town', in T. G. McGee (ed.) *Proletarianization in Asia and the Pacific*, Centre for Southeast Asian Studies Monograph, Monash University.
Robinson, S. (1982) 'A change from Nasi Lemak', *Business Times*, Singapore, 26 June.
Rocamora, Joel (1983) 'Japanese capital in Indonesia: benefactor or beneficiary?', *Southeast Asia Chronicle*, 88, 7–9 & 13.
Salaff, J. W. (1981) *Working Daughters of Hong Kong*, New York, Cambridge University Press.
—— and Wong, Aline (1982) 'Women's work; factory, family and social class in an industrializing order', paper presented at the Conference on Women in the Urban and Industrial Workforce, Southeast and East Asia, Manila, University of the Philippines.
Salih, Kamal (1981) 'Malaysia and the world system: a perspective essay on incorporation, social groups and the state', unpublished paper presented to Development Centre, OECD, Paris, 28–30.
—— (1983a) 'Urbanization in Malaysia. Impact of the new economic policy', paper prepared for Parliamentarian Seminar, Kuala Lumpur.
—— (1983b) 'Penang's development 1970–1980': a case study of planned growth and unplanned consequences', paper presented to International Seminar on Cheju Island Regional Development, Cheju, South Korea.
—— *et al.* (1982) *Progress Report Malaysian Case Study 'Urban Services for Young Workers in Penang'*, Penang, School of Social Sciences, University Sains.
Scott, J. (1976) *The Moral Economy of the Peasant*, New Haven, Yale University Press.
Silin, R. H. (1965) 'Trust and confidence in a Hong Kong wholesale vegetable market', unpublished M.A. thesis, Department of Anthropology, University of Hawaii.
Silin, R. H. (1972) 'Marketing and credit in a Hong Kong wholesale market', in W. E. Willmont (ed.) *Economic Organization in Chinese Society*, Stanford, Calif., Stanford University Press, 327–52.

Singh, M. H. (1979) 'Spatial dynamics in the growth and development of multinational corporations in Malaysia', unpublished paper presented to IGU Commission on Industrial Systems Symposium, Rotterdam.

—— and Choo, C. S. (1981) 'Spatial dynamics in the growth and development of multinational corporations in Malaysia', in E. E. I. Hamilton and G. J. R. Linge (eds) *Spatial Analysis, Industry and the Industrial Environment*, New York, John Wiley.

Smith, Wendy, A. (1982) 'The impact of Japanese foreign investment and management style on female industrial workers in Malaysia', paper presented to the Conference on Women in the Urban and Industrial Workforce, Southeast and East Asia, Manila, University of the Philippines.

Song, Byung-Nak (1978) 'Economic growth and rural-urban relations in Korea', Working Paper 78–04, Nagoya, United Nations Centre for Regional Development.

—— (1979) 'Regional policy for Korea's Fifty Year Plan 1982–86', paper prepared for United Nations Centre for Regional Development Seminar, Oct./Nov., Nagoya.

Taerssen, Van A. (1980) 'Location factors and linkages at the industrial estates of Malacca Town', *Research Notes and Discussion Paper, 16*, Institute of Southeast Asian Studies, Singapore.

Takeo, T. (1978) 'Free Trade Zones in Southeast Asia', *Monthly Review*, 29, 9.

Tse, F. Y. (1971) 'A preliminary report on the operation, size, daily turnover and gross profit variables of stall traders in Hong Kong', Hong Kong, Social Research Centre, The Chinese University.

Turner, J. F. C. (1968) 'Housing priorities settlement patterns, and urban development in modernizing countries', *Journal of the Institute of American Planners*, 34, 6, 354–63.

Ward, B. E. (1972) 'A small factory in Hong Kong: some aspects of its internal organization', in W. E. Willmont (ed.) *Economic Organization in Chinese Society*, Stanford, Calif., Stanford University Press, 353–85.

Watanabe, S. (1972) 'Exports and employment: the case of the Republic of Korea', *International Labour Review*, 106.

Westphal, L. E. (1982) 'The private sector as "principal engine" of development: Korea', *Dyason House Papers*, 8, 4, 7–16 (Victorian Branch of the Institute of International Affairs).

Wheelwright, E. L. (1965) *Industrialization in Malaysia*, Melbourne University Press.

Willmont, W. E. (ed.) (1972) *Economic Organization in Chinese Society*, Stanford, Calif., Stanford University Press.

Wong, Aline, K. (1979) 'The changing role and status of women in ASEAN', *Contemporary Southeast Asia*, 1, 2, 179–93.

Wong, J. (1970) 'The dynamics of the vegetable market in Hong Kong', The Hong Kong Institute of Social Research Seminar Paper (mimeo).

World Bank (IBRD) (1979) *World Development Indicators*, Washington, DC.

World Bank (1982) *World Development Report 1982*, New York, World Bank/Oxford University Press.

—— (1984) *World Development Report 1984* New York, World Bank/Oxford University Press.

Wu, Yuan-li and Wu, Chun-hsi (1980) *Economic Development in Southeast Asia: the Chinese Dimension*, Stanford, Calif., Hoover Institution Press, Stanford University.

Young, M. L. and Salih, K. (1982) 'Urbanization trends in Malaysia: some policy directions for the Eighties', unpublished paper, University Sains, Malaysia, Penang.

Index

Note: Sub-entries are in alphabetical order, except where chronological order is significant.

For Product Safety Concerns and Information please contact our EU representative GPSR@taylorandfrancis.com
Taylor & Francis Verlag GmbH, Kaufingerstraße 24, 80331 München, Germany

www.ingramcontent.com/pod-product-compliance
Lightning Source LLC
LaVergne TN
LVHW011118300926
849160LV00008B/434